RF and Microwave Microelectronics Packaging II

Ken Kuang • Rick Sturdivant
Editors

RF and Microwave Microelectronics Packaging II

Springer

Editors
Ken Kuang
Torrey Hills Technologies, LLC
San Diego, CA, USA

Rick Sturdivant
Azusa Pacific University
Azusa, California, USA

ISBN 978-3-319-84719-1 ISBN 978-3-319-51697-4 (eBook)
DOI 10.1007/978-3-319-51697-4

Softcover reprint of the hardcover 1st edition 2017

Printed on acid-free paper

This Springer imprint is published by Springer Nature
The registered company is Springer International Publishing AG
The registered company address is: Gewerbestrasse 11, 6330 Cham, Switzerland

From Ken Kuang:
To my family, Zheng, Simon and Andrew, you are my inspiration and support.

From Rick Sturdivant:
Dedicated to my parents, Jim and Linda Sturdivant.

Foreword

The question how to package an electronic component has become a key issue in RF system development. In many cases it decides about the "to be, or not to be". In the end, it is a combination of electrical, thermal and mechanical properties as well as fabrication cost which makes the difference. All these aspects need to be taken into account and this multi-disciplinary approach distinguishes packaging development from classical circuit design. Moreover, when packaging modules in the microwave and mm-wave range, there is no general solution which could be used as a standard overall. The best choice varies depending on frequency range, application, volume and cost limitations, just to name some of the most important boundary conditions. While in the low GHz range the situation is relatively mature and high-volume low-cost approaches are in routine use for the mobile communications market, in the mm-wave frequency range above 100 GHz the lack of cost-competitive and volume-compatible packaging solutions is still a big bottleneck in getting systems to the market.

So, writing a book addressing this extensive and multifaceted field is a major challenge. The authors have chosen their own approach and compiled a selection of expert contributions covering the full scope of topics, from modules to materials and from 3D transitions to thermal management. The reader should not expect an in-depth introduction to all of these fields, which would exceed the scope of a single book by far. Rather she/he will find in the different chapters articles on the various topics that help in getting familiar with the subject and explain typical state-of-the-art solutions. Thus, this book provides a good overview as well as useful hints for the practitioner and I am sure it will prove to be a worthwhile addition to the hand library of many microwave engineers.

Head of Microwave Department
Ferdinand-Braun-Institut, Leibniz-Institut
für Höchstfrequenztechnik
Berlin, Germany

Wolfgang Heinrich

Foreword

The past decade has been an exciting time in RF. We have seen mobile communication become a technology for the masses. This not only lead to the rise of new types of mobile phones: the smartphone like iPhones to Android handsets, it also started a wide range of standards from 3G to 4G, LTE, 4.5G and soon 5G. Every new standard adds complexity to the system. Not only because new standards mean more functions but also more challenges for the packaging engineer in putting all different standards in one handy package while reducing interference between them. The contribution of Nozad Karim gives some good examples.

Looking at the other side of the mobile communication line, the basestations, we see a similar proliferation of technology. Take for instance the RF Power amplifiers where basestation manufacturers are moving from traditional LDMOS transistors to GaN transistors. The higher power densities of these new semiconductor technologies require heat spreaders with better heat spreading capabilities. Standard CuW or CPC heat spreaders are not, anymore, fulfilling the requirements put on them by advanced amplifier designs. Managing the dissipated power of these amplifiers can be done by new materials as described by Dr. Wei Fan and Kevin Loutfy.

The mobile phone industry is one of the more visible areas where RF and microelectronics are used. This does not mean that there are no other applications for high frequency radio waves. Other interesting developments are the use of RF technology in non-traditional applications like RF Cooking (Microwaves), RF Lighting (using RF induced plasmas) and other technologies where RF Energy is used in other ways than for communication. Many of these new areas are more consumer oriented than traditional infrastructure or aerospace and defense electronics. This also has implications on the packaging. Cost becomes more prominent than ever. Also ease of use for customers not used to RF technology is an increasingly important aspect. This drive for lower cost can be found back in the transition from ceramics materials to laminates. It is also seen in new developments in modules and integration of complete RF systems into the package to help customers with less RF knowledge build a working system. Several of these developments are discussed in this book by Rick Sturdivant, amongst others.

Finally, I would not do justice to all developments in the more traditional RF packaging if I would not mention the progress in packaging of advanced RF systems like shielding, filter design and advanced thermal solutions. Dr. Min Tan and Dr. Jonathan Holmes give some interesting examples of these developments.

I'm very excited to see this successor to "RF and Microwave Microelectronics Packaging", bringing engineers in electronic packaging the latest developments in high frequency electronics. I hope the reader will enjoy the different articles from academics as well as industry research as much as I did.

Ampleon Semiconductors
Nijmegen, The Netherlands

Michel de Langen

Contents

Chapter 1
Introduction to Radio Frequency and Microwave Microelectronic Packaging

Rick Sturdivant

1.1 Introduction

Electronic packaging at radio frequency (RF) and microwave frequency has become an important part of engineering development for a wide variety of products. Mobile phones, for instance, contain a multitude components such as processors, RF amplifiers, antennas, and passive components into a compact and robust form factor. The processors and amplifiers can dissipate significant heat, which requires proper heat transfer design. In addition, the high-speed digital signal lines, microwave transmission lines, and interconnects must be designed to maintain the integrity of the signal. The antennas in mobile phones are often integrated into the case or circuit boards and radiate not only away from the phone but also into the phone circuits. This means that sensitive circuits in the phone must be designed to minimize coupling of the phone's wireless signals that can cause undesirable effects such as oscillations or resonances in the circuits. Passive components must be carefully chosen and implemented because their physical size is a significant fraction of a wavelength at the operating frequency. As a result, the passive components may no longer function as ideal capacitors, inductors, or resistors. Rather, they may have distributed effects that must be taken into account to achieve desired performance. For these reasons, and others, the task of electronic packaging is widely recognized as a critically important task for development of products operating at microwave and millimeter-wave frequencies.

This chapter is concerned with providing an introduction to the main issues and concerns encountered in developing products at radio and microwave frequencies. The focus is on presenting fundamental principles in three of the most important areas. This first is distributed effects. As mentioned earlier, as the frequency of operation of a circuit increases, the size of the circuit itself and/or its components become

R. Sturdivant (✉)
Azusa Pacific University, Azusa, California, USA
e-mail: ricksturdivant@gmail.com

K. Kuang, R. Sturdivant (eds.), *RF and Microwave Microelectronics Packaging II*,
DOI 10.1007/978-3-319-51697-4_1

an appreciable fraction of a wavelength. The result is that special care must be taken in detailed design. The second area is transmission lines, which are used to carry RF and microwave signals from one point in a circuit or system to a different point. Basic transmission line concepts are introduced, such as line impedance, propagation constant, group delay, and dispersion. The third is materials. There are a few important material characteristics that are particularly important when developing products at these frequencies, such as dielectric constant, dielectric loss tangent, coefficient of thermal expansion, and thermal conductivity. These material parameters are introduced and simple examples are given. These topics are addressed in detail in following chapters, but this chapter introduces these ideas and provides the fundamental concepts that the following chapters build upon.

1.2 Frequency Bands

As a matter of convenience and based on historical usage, the radio spectrum has been divided into bands. This is useful since practicing engineers find it helpful to refer to the frequency of operation for a system or component by its band of operation. Table 1.1 shows frequency band designators and nominal frequency range up to 300 GHz. For completeness, the table includes band designators down to 3 MHz.

It is useful to understand a few of the types of applications used for each of the frequency bands. Therefore, a partial listing of the types of applications for each frequency band is provided. At low frequencies, the high-frequency (HF) band has been used for amateur radio, radio frequency identification (RFID) (at 13.56 MHz), radar such as Doppler ocean wave radar [1], the AN/TPS-71 over-the-horizon radar

Table 1.1 Industry standard frequency band allocations

Band designator	Frequency range	Comments
HF	3–30 MHz	HF refers to high frequency
VHF	30–300 MHz	VHF refers to very high frequency
UHF	300–1000 MHz	UHF refers to ultra-high frequency
L	1–2 GHz	
S	2–4 GHz	
C	4–8 GHz	
X	8–12 GHz	
Ku	12–18 GHz	From the German *Kurz-unter* for under K-band
K	18–27 GHz	From the German *Kurz* which means short in English
Ka	27–40 GHz	From the German *Kurz*, Ka refers to above K
V	40–75 GHz	Also referred to as part of millimeter-wave frequencies
W	75–110 GHz	Also referred to as part of millimeter-wave frequencies
mm	110–300 GHz	mm refers to millimeter wave

[2], and other radio communication uses. The VHF band has been used for FM radio, television, amateur radio, land mobile, marine radio, and weather radio broadcasts. The UHF band is used by mobile phones, ZigBee (at 915 MHz and 868 MHz), RFID, amateur radio, fixed wireless, and television.

At microwave frequencies, L-band is used for radar, satellite communications, mobile phone, and amateur radio. S-band is also used by radio amateurs, radar applications in particular volume search radars utilize this band, mobile phones, RFID, Wi-Fi, ZigBee, Bluetooth, fixed wireless, and other applications in the popular 2.4 GHz ISM frequency allocation. C-band is used for satellite communication, military and weather radar [3], Wi-Fi, and other applications using the 5.8 GHz ISM band. X-band is used for military and weather radar applications, satellite applications (military), police radar, and wireless backhaul. Ku-band is used for satellite communication, satellite TV, police radar (in Europe), and wireless backhaul.

At millimeter-wave frequencies, Ka-band is used for satellite communication, military radar, and wireless backhaul, and is being investigated as part of new 5G systems [4]. V-band is planned for 60 GHz Wi-Fi, wireless backhaul, and satellite-to-satellite communication. W-band is used for automotive radar (at 77 GHz), satellite communication, passive imaging, wireless backhaul, and radar. The millimeter-band applications are emerging but include that same types of applications used at V- and W-bands.

These are a few examples of the applications for each of the frequency bands. There are many more that are not captured in this chapter since the list is not intended to be exhaustive.

1.3 Distributed Effects

The fundamental cause for many of the electrical challenges of packaging at RF and microwave frequency is that as frequency increases, the physical size of circuit elements become an appreciable fraction of a wavelength, or even larger than a wavelength. When this occurs, the electrical performance of components can change significantly. For instance, capacitors begin to have inductive effects and will even display an inductive–capacitive (LC) resonance. This effect even exists with simple surface mount resistors. This is in contrast to standard electric circuit theory that treats circuit elements as being much smaller than a wavelength so that distributed effects are neglected.

For example, consider the simple case of a 47 Ω surface mount resistor in a 0603 form factor as illustrated in Fig. 1.1a. At low frequencies, the resistor will be represented in simulation by a simple single element lumped resistor. However, as the operating frequency of the resistor increases, distributed effects affect the resistor's electrical performance. As a result, the simple single element circuit model is not sufficient.

The equivalent electrical circuit for high-frequency operation is shown in Fig. 1.1b. Note how the equivalent circuit has a capacitance in parallel with the resistor. This capacitance is partly due to the distributed effect of the resistor and to the stray

Fig. 1.1 A surface mount resistor (a) image for an 0603 package, and (b) the microwave frequency equivalent circuit that includes stray capacitance, Cp

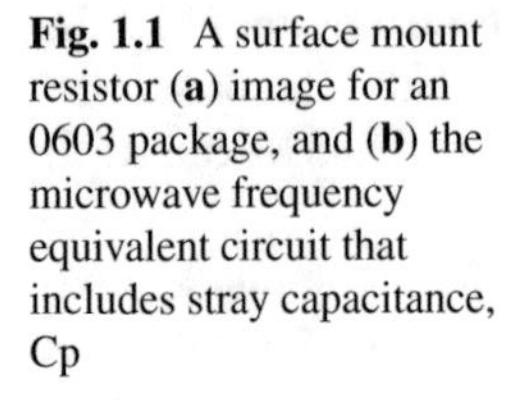

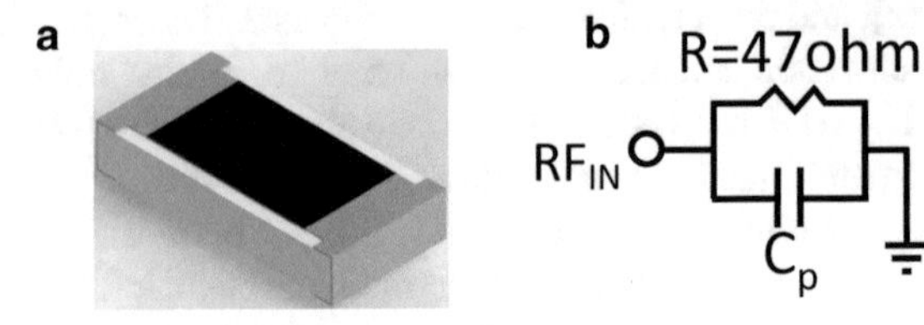

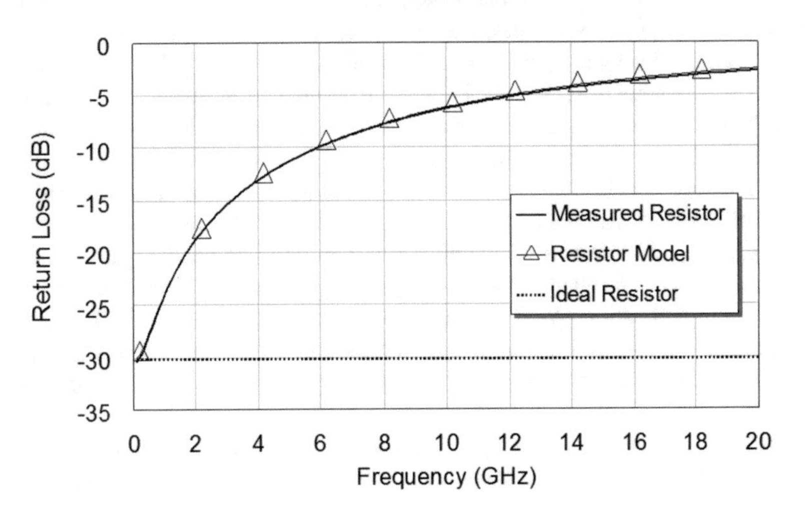

Fig. 1.2 Plot of the measured performance and model performance for a 47 Ω resistor in a 0603 case with Cp = 0.37 pF

capacitance of the resistor package itself. The distributed effect is approximated by a capacitance due to the capacitive effects of the printed resistor material relative to the ground reference. A transmission line could be used to model the effect, but a lumped capacitor was sufficient. The stray capacitance is due to the metal on the contact pads of the resistor. Figure 1.2 is a plot that shows three things about the resistor. First, it shows the measured resistor performance when it is connected shunt to ground. Second, it shows the simulated performance of the equivalent resistive–capacitive (RC) circuit. Note how the return loss increases as frequency increases, which is caused by the shunt capacitor, Cp. Third, the plot shows the simulated performance of an ideal 47 Ω resistor which is a flat line return loss of approximately −30 dB.

This example illustrates how distributed effects impact the performance of even simple passive elements, which is one reason packaging at microwave and millimeter-wave frequencies can be challenging.

The size of the 0603 resistor relative to the wavelength of operation can be found by first calculating the wavelength. The wavelength of a signal is calculated from the velocity and frequency of operation. The velocity depends upon the dielectric constant of the material. The wavelength can be calculated using

$$\lambda = \frac{v}{f} = \frac{v_0}{f\sqrt{\varepsilon_{r_{eff}}}} \quad (1.1)$$

where:

λ = wavelength (m).
ν = velocity of the signal in the material of the resistor body (m/s).
ν_o = velocity of light in a vacuum = 3 × 10^8 m/s.
$\varepsilon_{r_{eff}}$ = effective dielectric constant of the resistor body material (unitless).

Considering the 0603 resistor, its size is approximately 1.5 mm × 0.75 mm. Using Eq. (1.1) and assuming that the operating frequency is 10 GHz and that the dielectric constant of the resistor ceramic is 9.2 (for alumina ceramic), the wavelength can be found as

$$\lambda\Big|_{\substack{f=10GHz \\ e_{r_{eff}}=9.2}} = \frac{3\times10^8\left(\mathrm{m/s}\right)}{10\times10^9\left(Hz\right)\sqrt{9.2}} = 9.89mm.$$

This means a 0603 resistor is approximately 15% of a wavelength at 10 GHz or 30% of a wavelength at 20 GHz, which is enough to cause distributed effects.

1.4 Transmission Lines

Transmission lines are used to carry RF and microwave signals from one point to another with minimum loss and degradation. The signal propagates along the transmission line. A transmission line can be represented by a model as shown in Fig. 1.3. Note that the model elements are a series inductance and resistance with a shunt capacitance and conductance, which are actually distributed quantities defined as follows:

L = distributed inductance (Henry/m).
R = distributed resistance (Ohm/m).
C = distributed capacitance (Farad/m).
G = distributed conductance (Siemens/m).

In Fig. 1.3, the length Δz is very small compared to the wavelength. The characteristic impedance of the transmission line is given by

$$Z_0 = \sqrt{\frac{R + j\omega L}{G + j\omega C}} \quad (1.2)$$

The resistance, R, and the conductance, G, model the losses in the transmission line. The losses are caused by the resistance of the metal conductors that are part of

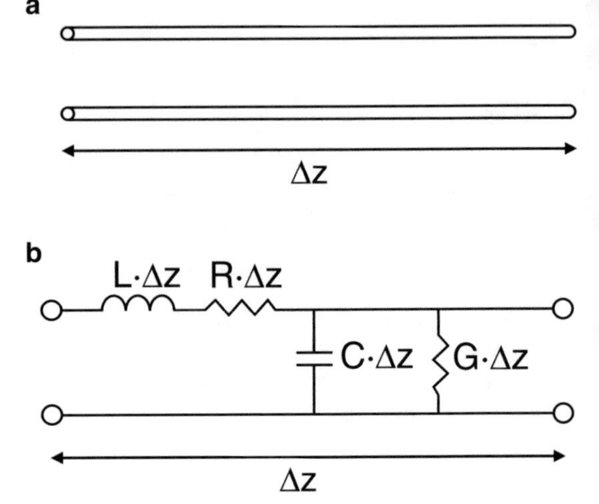

Fig. 1.3 Transmission line model

the transmission line, by the signal absorption that occurs in the dielectrics that are part of the transmission line, and by signal loss due to radiation. For the ideal case of a lossless transmission line where R and G are ignored, the line impedance is given by

$$Z_0 = \sqrt{\frac{j\omega L}{j\omega C}} = \sqrt{\frac{L}{C}} \tag{1.3}$$

It is convenient to think of Eq. (1.3) when thinking about transmission lines. This is because it is possible to consider how changes in the dimensions of a transmission line will change the capacitance and inductance and create a corresponding change in line impedance. For instance, increasing the line capacitance will decrease the line impedance.

As described earlier, the wavelength of a signal is a function of the frequency and the effective dielectric constant. The propagation constant is given by

$$\beta = \frac{2\pi}{\lambda} = \frac{2\pi f}{v} = \frac{2\pi f\sqrt{\varepsilon_{r_{eff}}}}{v_0} \tag{1.4}$$

This is an important parameter since the delay of a signal as it propagates along a transmission line of length, l, is given by

$$\text{Signal Delay} = \text{Phase Shift} = \beta l = \frac{2\pi}{\lambda}\cdot l = \frac{2\pi f}{v}\cdot l = \omega\frac{l}{v} \tag{1.5}$$

The signal delay is actually a phase shift of the signal and is measured in radian angle. In other words, a signal traveling along a transmission line will experience a phase delay or phase shift.

1.5 Commonly Used Transmission Lines

In this section, four commonly used transmission line types are described. For each transmission line, the signal line is shown as well as the ground reference and dielectric support. Together, these physical features constitute the transmission line.

One of most common transmission line types is microstrip and it is shown in Fig. 1.4a. Note how the signal line exists on top of a dielectric support substrate. The dielectric has a ground plane underneath it. One of the reasons for the wide use of microstrip is that it is easy to obtain access to the signal line on top of the dielectric substrate and fabrication of microstrip is easily accommodated by most circuit board fabrication processes. This means that it is easy to connect to surface mount components and to integrate circuits such as couplers and filters.

Another transmission line type (often used in multilayer modules and packages) is stripline, which is illustrated in Fig. 1.4b. Note that the signal line is buried in the dielectric substrate below a top ground plane and above a lower ground plane. One of the advantages of this approach is that it can achieve good isolation between adjacent circuits, can be used to route RF signal lines below other circuits, and can be used to bury circuits such as filters to increase the total circuit density of a module or package. It is important that the top and bottom ground planes are connected using vias to avoid the excitation of undesired modes. The location, spacing, and pitch of the vias must be carefully chosen [5]. Also, vias between ground planes are necessary to achieve isolation between circuits [6]. When properly designed, stripline can be a useful transmission line for both ceramic and laminate packaging.

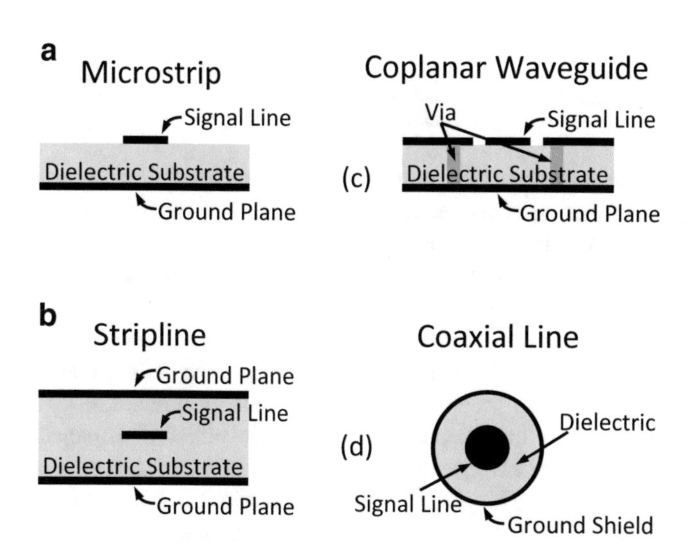

Fig. 1.4 Four common transmission line types: (**a**) microstrip, (**b**) stripline, (**c**) conductor backed coplanar waveguide, and (**d**) coaxial line

Conductor backed coplanar waveguide (CBCPW) is useful for circuits that require access to the ground reference on the topside of the dielectric substrate. It is illustrated in Fig. 1.4c. It is also useful for circuits that require improved isolation over what can be achieved with microstrip. One drawback is CBCPW requires the use of a via structure that connects the topside ground planes to the lower ground plane to avoid signal leakage, resonances, and coupling issues [7, 8]. This transmission line finds wide use since it can easily interface with surface mount components and is nearly as easily as microstrip line to fabricate using standard processes.

Coaxial transmission line may be the most familiar since it is very common for television connectivity. It is illustrated in Fig. 1.4d. A major advantage of coax is that it has very good isolation. A drawback is that it is difficult to use coax in planar circuit designs.

1.6 Dispersion in Transmission Lines

Returning to the discussion of transmission line signal propagation, an important parameter used to characterize transmission lines and some components such as filters is group delay. It is the rate of change of the phase shift with respect to ω, which can be written as

$$\text{Group Delay} = \frac{d\phi}{d\omega} \tag{1.6}$$

Combing Eq. (1.5) with Eq. (1.6) and performing the differentiation, it is simple to show that

$$\text{Group Delay} = -\frac{d\phi}{d\omega} = -\frac{d\left(\omega l / v\right)}{\omega} = -\frac{l}{v} = -\frac{l}{v_0}\sqrt{\varepsilon_{\mathrm{r}_{\mathit{eff}}}} \tag{1.7}$$

It is important to realize that, in general, the effective dielectric constant is considered to be a function of the operating frequency. For some transmission lines, it is a weak function of frequency, but for others, it is a strong function of frequency. Those transmission lines that have a strong dependence of the effective dielectric constant with frequency are said to be dispersive.

Dispersion in transmission lines can have an important effect on the performance of packaging in RF and microwave circuits. This is especially a concern for wide band circuits such as high-speed digital signals which must maintain signal integrity over very wide bandwidths. Dispersion describes the effect of propagation constant not varying linearly with frequency. This in turn causes the group delay to not be constant over frequency. This effect is illustrated in Fig. 1.5. Note how the microstrip line group delay varies over frequency due to dispersion but is constant for coax since it is non-dispersive over a wide frequency range.

The result is that some frequency components of a wide band signal will be delayed relative to other frequency components, which can cause jitter, eye closing,

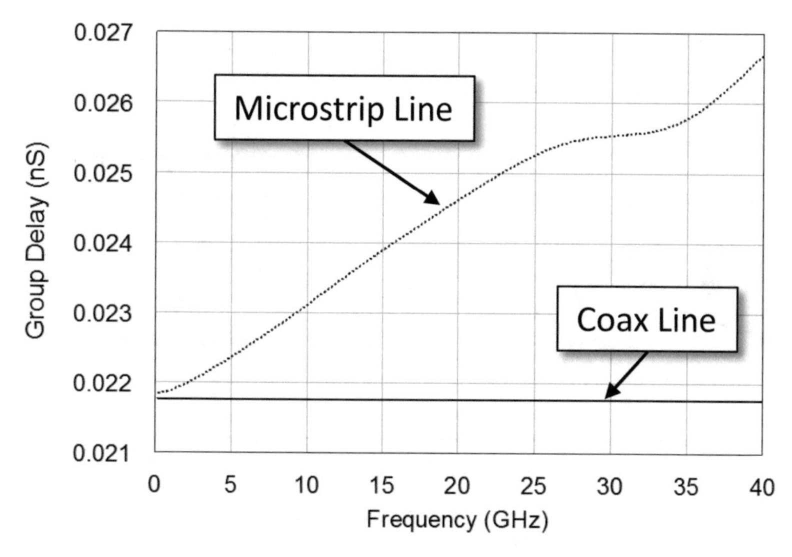

Fig. 1.5 Dispersion causes the group delay of a transmission line to not be constant as a function of frequency (microstrip: line width = substrate thickness = 0.635 mm, $\varepsilon_r = 9.8$, coax: outer conductor diameter = 1.905 mm, inner conductor diameter = 0.222 mm, $\varepsilon_r = 6.6$, l = 2.5 mm)

overshoot, undershoot, and rail widening in the eye diagram. Figure 1.6a and b shows the effects of dispersion in transmission lines for a 10 Gb/S digital signal that propagates along a coax transmission line and microstrip transmission line. Since coax is a zero dispersion transmission line when operated below the propagation of higher order modes, the eye diagram shows clean eye opening with very small amount of overshoot and rail widening. The case for microstrip, on the other hand, shows overshoot and eye closing that are caused by dispersion. Of course the particular features of the microstrip line were chosen, in this case, to exhibit dispersion for illustration purposes.

To minimize dispersion, the thickness of the substrate should be chosen to be lesser than approximately 10% of a wavelength at the operating frequency in the material of the substrate. This can be written as

$$\frac{h}{\lambda} = \frac{hf\sqrt{\varepsilon_r}}{v_0} < 0.1 \tag{1.8}$$

1.7 Dielectric and Substrate Materials

There are many different types of nonmetallic substrates used for microelectronic packaging at RF and microwave frequencies. For simplicity sake, we will consider three. The first is laminate substrates. This includes the familiar FR-4 materials, polytetrafluoroethylene (PTFE) that are used for microwave laminates, and hybrid

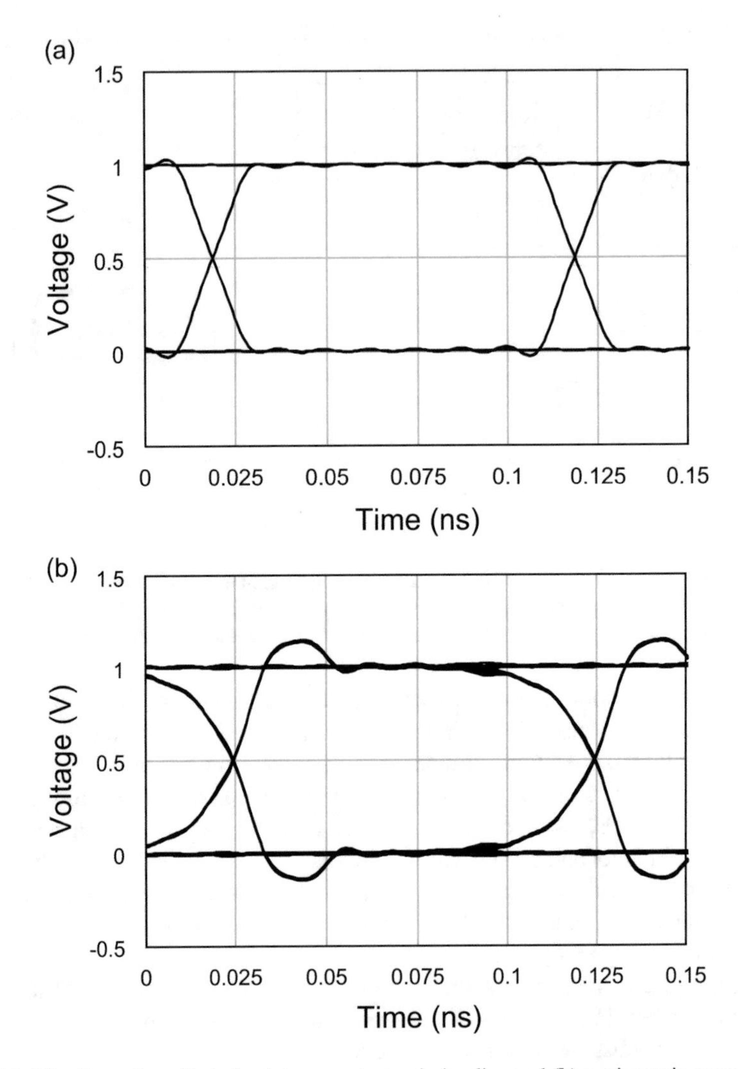

Fig. 1.6 The dispersion effects for (**a**) a coax transmission line and (**b**) a microstrip transmission line (dimensions that are same as for Fig. 1.5, but $l = 12.7$ mm)

materials. The common thread for laminates is that they are fabricated using the same fabrication method, which is the lamination of discrete layers together under pressure and temperature. There are core layers and prepreg (glue) layers that are patterned with copper using an etching process, treated, stacked, and then laminated at 340–400 °C for up to 90 min. The processing occurs in large panels which are drilled, plated, and surface finished before being cut to singulate the individual circuits boards. Lamination is a low-cost method to fabrication substrates at RF and microwave frequencies.

The second substrate material discussed is ceramics. Broadly speaking, ceramics are fabricated using four different technologies (other techniques are used, but these four provide a good overview of ceramics for RF and microwave packaging). The four are as follows:

1. *Thin Film Ceramics*: These ceramic substrates are fabricated using plates of ceramic that are finish polished normally to a high gloss. By far, the most common ceramic base material for thin film is alumina (Al_2O_3) with high purity level such as 99% or better. The ceramic is often pre-machined with a laser to form vias and other features. Metallization is deposited in layers and plated. The first layer deposited is normally a TiW layer that is 100–500 Å thick. This layer acts as a 'glue' so that subsequent metal layers will stick to the ceramic. Next, nickel is sputtered to a thickness of 1000–3000 Å. It acts as a barrier layer for additional layers. A gold layer is then deposited and plated to the desired thickness. For RF and microwave applications, the gold thickness is normally 1–3 μm (40–125 μin) thick. The desired circuit is etched out of the deposited and platted metal. Since the metal patterns are formed using an etching away method, thin film is said to be a subtractive process since material is 'subtracted' from the substrate to created the desired pattern. By using a glue layer that is electrically resistive, thin film resistors can be formed. Also, multiple layers of conductors can be created by depositing and processing dielectric layers and subsequent additional metal layers. A major benefit of thin film is that it can realize very fine lines and spaces down to 25 μm or less, which is important for some circuits especially for millimeter-wave applications.
2. *Thick Film Ceramics*: Thick film circuits also use plates of alumina. However, it is common for lower purity alumina such as 96% to be used and lower surface finish compared to thin film. The substrates are first machined with a laser to create vias, cutouts, and alignment features. The plates are then screen printed with metallization and fired in a belt furnace at temperatures between 500 and 1000 °C depending upon the materials being used. Layers of dielectric are also printed in sequence after metal prints to create a multilayer substrate. Resistors are also printed and normally 3–4 ink resistivity values can be used so that a large range of resistance values can be fabricated. Normal line and gap resolution for printed thick film is approximately 100–150 μm. Advanced thick film processes include etched metallization so that fine lines and spaces down to 25–35 μm can be fabricated. Since the layers are added to the substrate, thick film is considered an additive process.
3. *Low-Temperature Co-Fired Ceramics (LTCC)*: This process uses multiple layers of ceramic 'green' tape that is punched with holes, hole filled with metal ink, print metalized, dried, stacked, and then fired to create a monolithic piece of ceramic with internal and external metallization patterns. LTCC is attractive for microwave circuits since manufacturers have created tape materials) that are low loss and can use noble metals such as gold, silver, and even copper.
4. *High-Temperature Co-Fired Ceramics (HTCC)*: This is also a multilayer process similar to LTCC except the firing temperature can be as high as 1600 °C. The most popular material for HTCC is alumina and it is common to use 92% purity material. HTCC can also be fabricated using aluminum nitride (AlN). An important characteristic of HTCC is that since the firing temperature is so high, only refractory metals are used such as tungsten (W) or molybdenum (Mo).

The third material introduced in this section is semiconductors. For low-frequency applications, the most common semiconductor material is silicon (Si). At RF and microwave frequencies, it is common to use a much wider variety of semiconductor materials such as gallium arsenide (GaAs), silicon germanium (SiGe), gallium nitride (GaN), silicon carbide (SiC), and indium phosphide (InP). Each of these semiconductor materials has features that make it attractive, but all share the fact that they have superior electron transport performance so that they can be used to fabricate high-speed transistors.

An important dielectric material property at RF and microwave frequencies is dielectric constant. It is a measure of how a material reacts at the atomic level to an electric field. The electric field causes atomic polarization and a displacement vector is created in the material. From Maxwell's equations, we know that

$$D = \varepsilon_0 E + P = \varepsilon_0 E + \varepsilon_0 \chi_e = \varepsilon_0 \left(1 + \chi_e\right) E = \varepsilon_0 \varepsilon_r E \tag{1.9}$$

where:

P = dipole moment density (or polarization density) (coulomb/m^2).
E = electric field (V/m).
D = electric field displacement (coulomb/m^2).
ε_o = permittivity of free space (F/m).
ε_r = relative permittivity or relative dielectric constant (unitless).

Therefore, the electric field displacement is proportional to the dielectric constant of the material. Most manufacturers of materials provide dielectric constant as part of their data sheets. Most microwave dielectric materials have a dielectric constant that is from approximately 2 to 10, though there are high dielectric constant materials.

Loss tangent is also an important material parameter since it indicates the signal losses that can be expected in the transmission lines. Dielectric materials that are good for microwave circuit performance have low loss tangent typically below 0.005.

1.8 Skin Depth

An electromagnetic wave generates a current when it impinges on a metal surface, and the current will decay as the signal penetrates the metal. This effect is illustrated in Fig. 1.7. The current decays in the material according to

$$J_x\left(z\right) = J_0 e^{-z/\delta} \tag{1.10}$$

where:

J_o = Current at the surface of the metal.
z = distance into the metal (m).
δ = skin depth (m).

Skin depth is calculated using

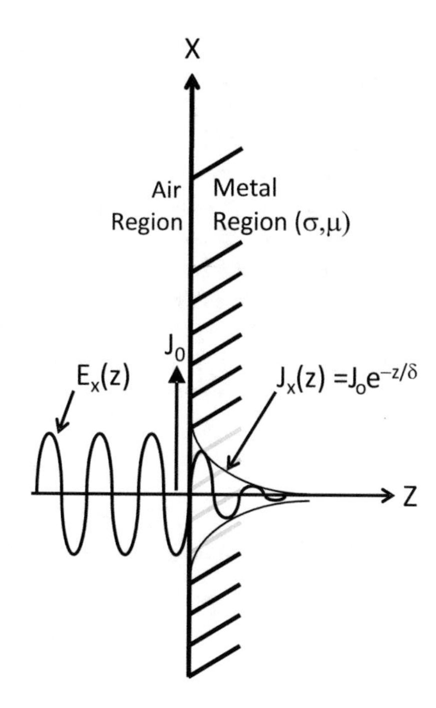

Fig. 1.7 Skin effect causes the RF signal to decay as it penetrates a metal surface (used with permission from 9)

$$\delta = \frac{1}{\sqrt{\pi \mu \sigma f}} \tag{1.11}$$

where:

μ = permeability = $\mu_0 \mu_r$ (H/m).
μ_o = permeability of free space (H/m).
μ_r = relative permeability (unitless).
σ = the direction of signal penetration into the metal (m).
δ = skin depth (m).

Skin effect is important since it tells us that the RF and microwave signals only penetrate a small distance into metals. This impacts the metallic losses that RF and microwave signals experience. It also shows that plating high-conductivity metals over lower-conductivity metals only needs to be a thin layer.

1.9 Thermal Conductivity, Electrical Conductivity, and Thermal Expansion

Materials are characterized by many different parameters, but the three that will be considered in this section are thermal conductivity, electrical conductivity, and thermal expansion. The reason they are discussed in this section is that they are normally among the first parameters a packaging engineers investigates for a new material (along with dielectric constant and loss tangent that were introduced in the last section of this chapter). Thermal conductivity describes the ability of a material to conduct heat and is applicable not only to metals but also to most every other material including dielectrics and semiconductor materials. Electrical conductivity is the ability of a material to conduct electrical current. Thermal expansion describes how a material expands or contracts as temperature changes.

The units of thermal conductivity are normally given as watt/meter·kelvin (W/mK). Fourier's Heat Transfer Law tells us that heat energy, Q, that is normal to an area, A, can be calculated from the thermal conductivity, k, multiplied by the change in temperature, ΔT, over a change in distance, ΔZ. This situation is illustrated in Fig. 1.8 and can be written as

$$Q = -kA\frac{\Delta T}{\Delta Z} \tag{1.12}$$

If ΔZ is taken to be the thickness, t, then this can be re-written to solve for ΔT as

$$\Delta T = T_2 - T_1 = \frac{Qt}{kA} \tag{1.13}$$

As can be seen, the temperature rise increases as the magnitude of heat source is increased and thickness is increased. The temperature rise decreases as the thermal conductivity increases and the area increases.

The electrical resistance of a material describes the ability of the material to inhibit the movement of electric current through it. Figure 1.9 illustrates the resistance from point R_2 to R_1 in a material with a resistivity, ρ (Ω·m), and area, $A = X \cdot Y$. The resistance can be written as

$$R = R_2 - R_1 = \rho\frac{\Delta Z}{X \cdot Y} = \rho\frac{l}{A} \tag{1.14}$$

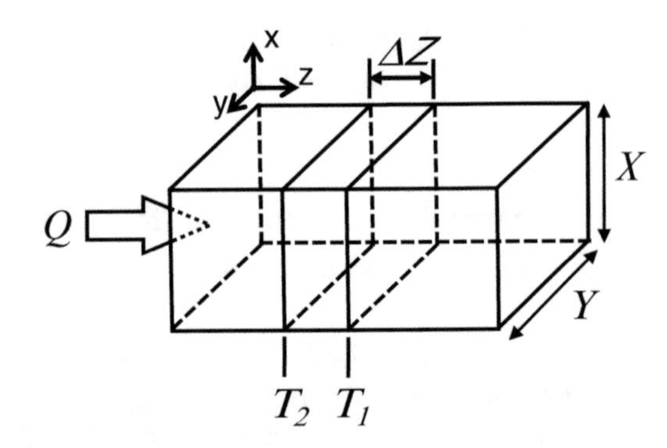

Fig. 1.8 Fourier's Heat Transfer Law can be used to describe how the temperature rise in a material due to a heat source is a function of area and thermal conductivity of the material

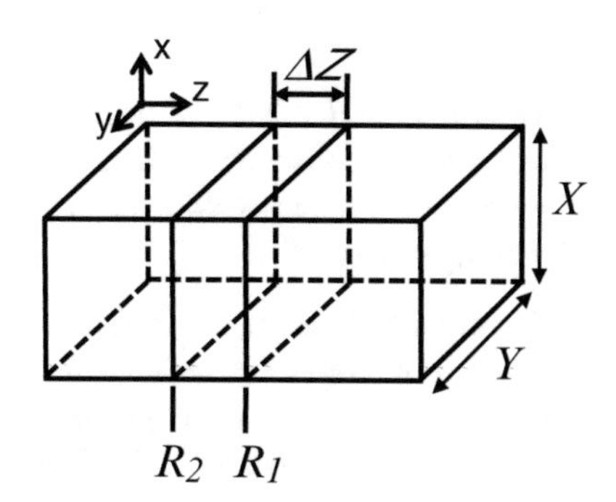

Fig. 1.9 Electrical resistance of a material

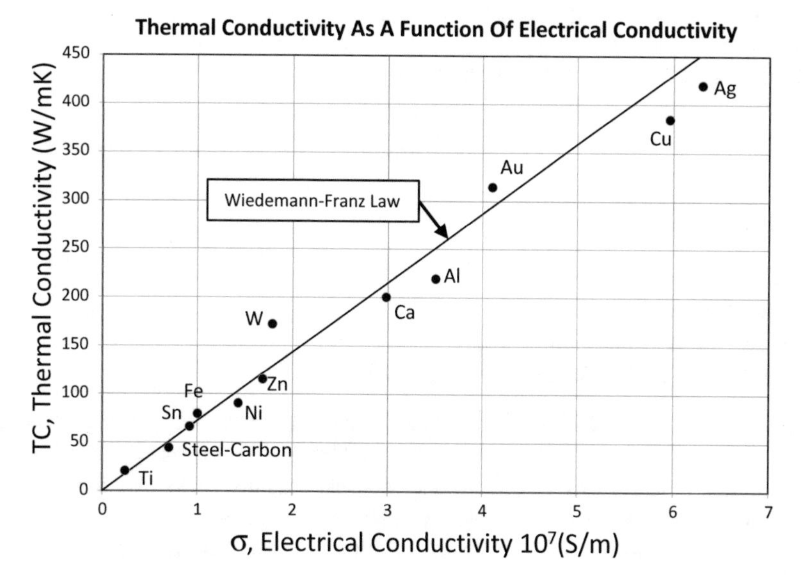

Fig. 1.10 Thermal conductivity is plotted as a function of electrical conductivity, which shows that metals obey the Wiedemann–Franz law (used with permission from 9)

Resistance is specified in units of Ohms. This tells us that the resistance increases as the length of the resistive material increases and as the resistivity increases. The resistance decreases as the area increases. Many materials are described by their electrical conductivity, σ, which is specified in Siemens/m or mho/m.

It is interesting and instructive to recognize that thermal conductivity and electrical conductivity in metals result from the existence of free electrons in the atomic structure. As a result, there is a relationship between them called the Wiedemann–Franz Law, which is illustrated in Fig. 1.10. It states that the ratio of thermal conductivity due to electron processes to electrical conductivity is proportional to temperature and is the same for all metals according to the relationship

$$\frac{k}{\sigma} = LT. \tag{1.15}$$

where:

k = thermal conductivity (W/mK).
σ = electrical conductivity (Siemens).
L = Lorenz number = 2.44×10^{-8} (WΩ/K^2)
T = Temperature (°C).

Thermal expansion is a measure of how a material changes size as temperature changes. For simplicity, we consider linear thermal expansion, though the principle holds true too for volume expansion as a function of temperature. Consider the linear bar of material in Fig. 1.11. Note how the material has an initial length of L, but as temperature changes by ΔT, the size has increased to $L + \Delta L$. If effects of pressure on the material are ignored, and the expansion rate is assumed to be constant as a function of temperature, then the relationship of expansion of the material as a function of temperature can be written as

$$\frac{\Delta L}{L} = \alpha \cdot \Delta T \tag{1.16}$$

where:

ΔL = change in length (m).
L = initial length (m).
ΔT = change in temperature (°C).
α = coefficient of thermal expansion (ppm/°C).

An important application of thermal expansion for packaging is when dissimilar materials are rigidly attached. Dissimilar materials that are rigidly attached may expand at different rates, which introduce stress, strains, and other loading. This can lead to reduced system reliability. The packaging engineer must take these forces into account to design material solutions that achieve long-term reliability goals.

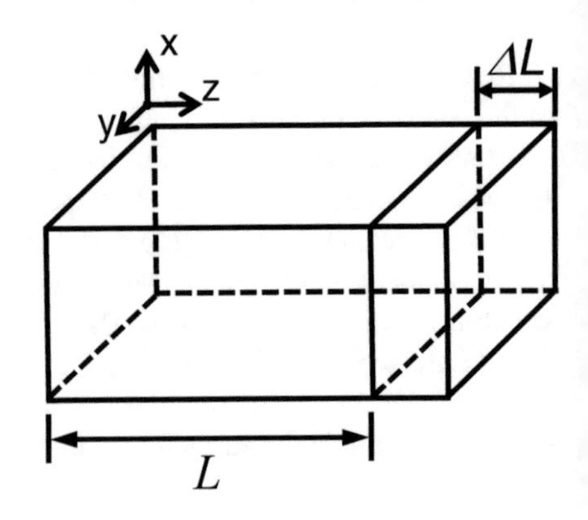

Fig. 1.11 Thermal expansion due to temperature changes

1.10 Chapter Conclusions

This chapter is an introduction to packaging at RF and microwave frequencies. It started with a review of the various operating and bands general applications. Distributed effect was introduced as the cause for many of the challenges of packaging at RF and microwave frequencies. In addition, basic concepts for transmission lines were described and four different transmission line types were reviewed. An important parameter for transmission lines is dispersion. It was described and its effect was illustrated with eye diagrams. Dielectrics such as laminated and ceramic substrate processing types were presented. It was shown that skin depth effect means that the RF current on a transmission line penetrates a small distance into the metal. Finally, thermal effects were introduced, such as thermal conductivity and expansion. These basic concepts will make the following chapters easier to understand and apply.

References

1. http://cordc.ucsd.edu/projects/mapping/documents/principles.php
2. Thomason, J.F. 2003. Development of over-the-horizon radar in the United States. In *Proceedings of IEEE international conference on radar*, September 2003, 599–601.
3. Marzano, F.S., and F. Vulpiani. 2004. Rain field and reflectivity vertical profile reconstruction from C-band volumetric data. *IEEE Transactions on Geoscience and Remote Sensing* 42(5): 44–48.
4. Gupta, A., and R.K. Jha. 2015. A survey of 5G network: Architecture and emerging technologies. In *IEEE access*, July 28, 2015, 1206–1232
5. Ponchak, G.E., D. Chen, J.G. Yook, and L.P.B. Katehi. 1998. Characterization of plated via hole fences for isolation between stripline circuits in LTCC packages. In *IEEE MTT-S international microwave symposium*, June 1998, Baltimore, MD, 1831–1834.
6. Gipprich, J., and D. Stevens. 2001. A new via fence structure for crosstalk reduction in high density stripline packages. In *IEEE MTT-S international microwave symposium*, May 2001, Phoenix, AZ, 1719–1722.
7. Liu, Y., and T. Itoh. 1994. Control of leakage in multilayered conductor-backed coplanar structures. In *IEEE MTT-S international microwave symposium*, May 1994, San Diego, CA, 141–144.
8. Assisi, A.H. 2009. Applying electromagnetic simulation to optimize via hole separation in conductor-backed coplanar waveguide. In *26th national radio science conference*, March 2009, Cairo, Egypt, 1–8.
9. Sturdivant, R. 2013. *Microwave and millimeter-wave electronic packaging*, 59. Norwood, MA: Artech House.

Chapter 2
Packaging of Transmit/Receive Modules

Rick Sturdivant

2.1 Introduction to Packaging of Transmit/Receive Modules

Transmit/Receive (T/R) modules were initially developed as the key component in phased array radar systems, which allowed the antenna beam to be scanned electronically. This was an important innovation over mechanically scanned arrays since it improved reliability, decreased beam scan time, and allowed other functionality. T/R modules are now used or are planned for use in satellites, wireless backhaul communication, mobile phones, Wi-Fi, and 5G mobile communication systems. As a result, development of T/R modules is expanding and shows no signs of slowing. A key to successful T/R modules is the use of the correct electronic packaging.

This chapter is divided into two main parts. The first part briefly describes phased arrays, gives an example T/R module block diagram, and discusses the major components of T/R modules such as integrated circuits, antenna elements, circulators/switches, beam forming network, and cooling sub-system. The first part concludes with examples of how phased arrays can be used with T/R modules in cellular base stations, Wi-Fi indoor location systems, 60 GHz Wi-Fi, and millimeter-wave point-to-point systems.

The second part of the chapter describes three types of packaging for T/R modules. They are a brick module, a tile array module, and a panel array. The particular challenges of each are given along with the typical solutions employed. This chapter ends with a discussion of integrated circuit/wafer level packaging T/R modules and a summary and conclusions.

R. Sturdivant (✉)
Azusa Pacific University, Azusa, California, USA
e-mail: ricksturdivant@gmail.com

K. Kuang, R. Sturdivant (eds.), *RF and Microwave Microelectronics Packaging II*,
DOI 10.1007/978-3-319-51697-4_2

2.1.1 Active Electronically Scanned Arrays

Active electronically scanned arrays (AESAs) scan their antenna beam electronically using phase shifters. This is illustrated in Fig. 2.1, which shows a simplified block diagram of a portion of an AESA. Note how phase shifters exist at each antenna element, which is generally true for most configurations. Normally, the phase shifter is implemented using an integrated circuit with digital control of the phase shifter, though it is possible to use analog control of the phase shift. Each phase state is independently controlled so that a continuous range of phase can be realized. For instance, an ideal block diagram of a 3-bit phase shifter is shown in Fig. 2.2. Note that the maximum phase state that can be achieved for a 3-bit phase shifter is

$$Max_{\text{Phase}} = 360° - LSB_{\text{Phase}} = 360° - 45° = 315°$$

Therefore, the range of phase is from 0° to 315° in 45° steps for a 3-bit phase shifter. This also means that an array constructed with 3-bit phase shifters will have eight distinct beam positions as the beam is electronically scanned. Though the phase shifter is an important component, it is only one of the parts required for T/R module functionality.

2.1.2 T/R Module Block Diagram

A simplified block diagram of a T/R module is shown in Fig. 2.3. Note how there is a circulator that connects the transmit and the receive signal paths to a common antenna. Some systems use a single-pole, double-throw switch for the antenna port connection. For the receive path the first component is the limiter. It is followed by

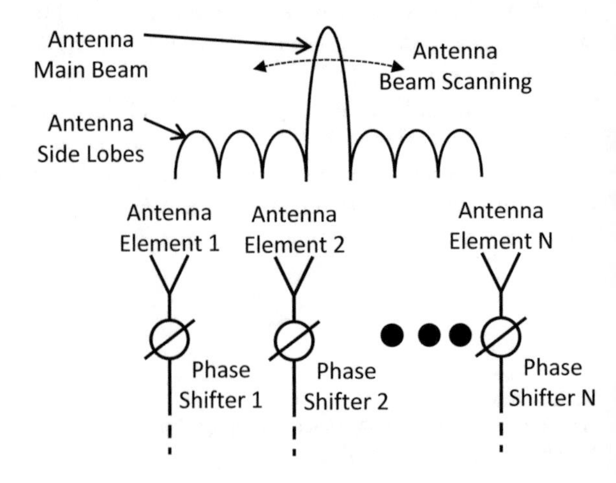

Fig. 2.1 A phased array has multiple antenna elements with independent phased control at each element in the array, which enables beam steering

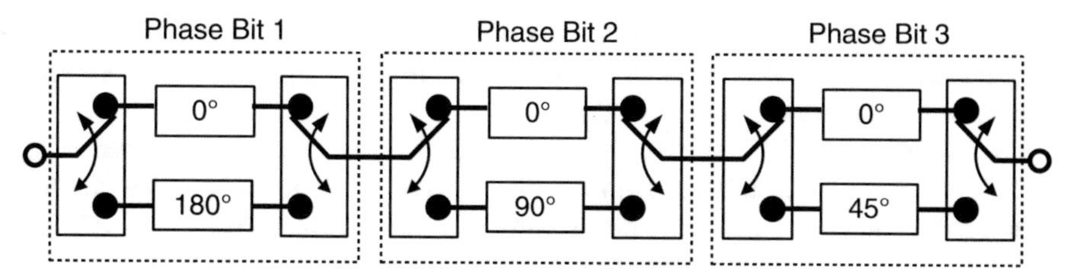

Fig. 2.2 Simplified block diagram of a 3-bit phase shifter using switches

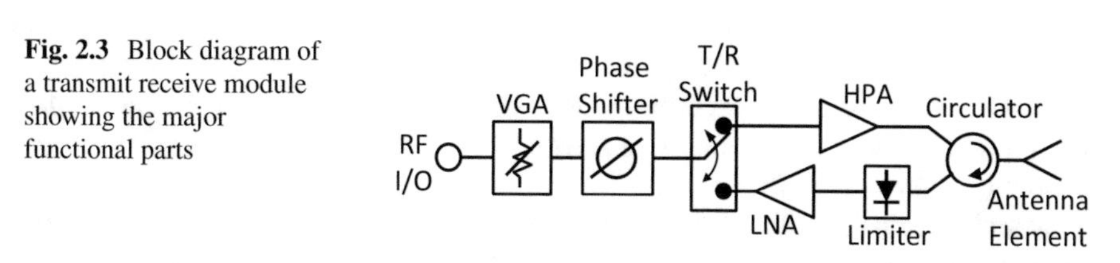

Fig. 2.3 Block diagram of a transmit receive module showing the major functional parts

the low-noise amplifier (LNA) which sets the noise figure of the receive path. This connects to an additional T/R switch that routes the receive signal to the phase shifter and variable gain amplifier and out the module input/output (I/O) port. On transmit, the signal enters the common port and is routed to phase shifter and variable gain amplifier. The T/R switch connects to the driver, which amplifies the signal to the level required by the high-power amplifier (HPA). Each of the components in the T/R modules is described in more detail in the following text.

High-Power Amplifier (HPA): It amplifies the transmit signal to the required output power. Communication systems operate the high-power amplifier in its linear range to minimize distortion of the signal. This typically means that the amplifier is operated below its compression point by 3–6 dB, and pre-distortion is often used to provide additional linearization. A good discussion of linear HPA operation can be found in [1]. Radar systems, on the other hand, often operate with the HPA fully saturated to achieve the highest RF output power possible.

Circulator: The circulator connects the transmit and receive portions together with the antenna. It is a non-reciprocal three-port device that allows one port to be RF electrically isolated from the other ports. Circulators are normally fabricated using ferrite substrates and magnets.

Low-Noise Amplifier (LNA): Amplifies the receive signal while adding minimal noise to the desired signal. The LNA is sensitive to input power and can actually be permanently damaged by a high signal level. The typical maximum input power is in the range of 0–25 dBm depending upon the design and semiconductor material used.

Limiter: The limiter is used to protect the LNA from permanent damage that can occur from signals that leak from the HPA or by signals that enter the antenna and travel past the circulator to the LNA. Limiters are typically fabricated using diodes such as PIN or Schottky. PIN diodes use an undoped intrinsic semiconductor between a p-type and an n-type doped region, which improves switching and limiting performance. As described in [2], PIN diode limiters can achieve fast limiting response and power clamping function.

T/R Switch: The T/R switch can exist in several locations in a T/R module. First, a T/R switch can be used instead of a circulator to provide connection of the transmit and receive function to the antenna. Second, a switch is often used in conjunction with a limiter to provide an additional level of protection for the LNA. Third, a switch is used to route the signals, so only one phase shifter and one variable gain amplifier are required for each T/R module.

Phase Shifter: The phase shifter provides the function that steers the antenna beam. As mentioned earlier, they are often realized using discrete phase bits connected in series. Alternative configurations use a vector modulator, analog control circuit (such as a varactor diode), or ferrite for the phase shifter. Phase shifters are normally implemented in 4–6 bits.

Variable Gain Amplifier: It is common for each element in the phased array to use variable gain amplifiers to perform amplitude calibration and antenna beam shaping.

2.2 Systems Using T/R Modules

The following sections describe how T/R modules are used or are planned for use in a variety of applications and describe some of the packaging methods used. Communication systems are considered first and include cellular base station, Wi-Fi-based indoor location systems, 60 GHz Wi-Fi high data rate connectivity, and millimeter-wave point-to-point systems. Military radar applications of transmit receive modules are also described. These examples show the variety of packaging methods employed for T/R modules.

2.3 T/R Modules in Communication Systems

2.3.1 Cellular Base Stations

Cellular base stations use antenna arrays that are installed on towers. Normally, the antennas are line arrays that are one or two elements by five to ten elements, depending upon the operating frequency and size requirement. The antennas are typically down tilted for efficient focus of the antenna on user locations. Normally, this down tilt is achieved by mechanically tilting the antenna, but this requires a manual modification of the antenna. Other methods use an electromechanical circuit that changes the phase of the signal to each antenna element in the line array to cause down tilt of the antenna beam. An alternative is to use phase shifters at each element in the array to achieve fully electronic beam steering that can be remotely controlled. The challenge is the phase shifters must handle the high power levels for base stations and, in most cases, need to perform beam steering while fully energized.

2.3.2 Wi-Fi Indoor Location Systems Using Phased Arrays

An application of phased arrays is used for indoor location systems. In this application, the indoor infrastructure uses multiple phased arrays operating in the Wi-Fi bands. This approach is illustrated in Fig. 2.4, which shows users in an indoor location with several phased arrays. The phased arrays scan to determine the position of mobile phones in the location. One approach uses multiple phased arrays, each measuring received signal strength (RSS) and location angle. Since each received signal from the users is unique (based upon SSID), it is possible to obtain location by advanced triangulation methods [3]. An example system is presented in [4], and it used four-element and eight-element linear arrays of antenna at 2.4 GHz. The analysis compared the linear array using a reflector and without a reflector. Another system is described in [5], which uses a single access point with multiple antennas arranged in a line array spaced at half a wavelength in the 5 GHz Wi-Fi band. The angle of arrival and time of arrival are used to compute the location of users within an environment. In another approach described in [6], a phased array operating in the 2.4 GHz band was used to locate users based upon RSS. It was found that accuracy was improved as the number of elements in the array increased. Also, improvement

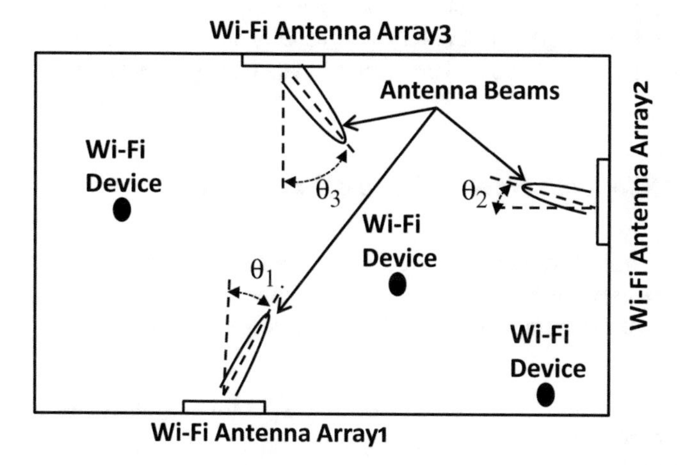

Fig. 2.4 Illustration of how Wi-Fi phased arrays can be used in an indoor location system with Wi-Fi-enabled devices

in location accuracy was increased by using the position found from antenna scanning and from the propagation delay (as in Radar). Phased arrays can be used for indoor location systems.

2.3.3 60 GHz Wi-Fi

The 802.11ad standard, which is also called Wi-GiG (Wireless Gigabit Alliance), will provide high-speed connectivity to mobile devices, televisions, computer displays, peripheral connectivity, peer-to-peer data transfer, and high-speed local area networks (LAN). The key capability for these systems are phased arrays and transmit/receive modules. Early work on the integrated circuit chip sets for these applications included phased arrays and was shown in [7–9]. The packaging used and proposed in these early solutions included low-cost laminate solutions with integrated antennas as shown in [8] and included methods that use low-cost FR-4 as the base laminate with liquid crystal polymer top layers and the use of multi-sector phased arrays for extended azimuth and elevation coverage [10]. The laminate material stack up for this type of approach is illustrated in Fig. 2.5. In these examples, the transmit receive module is fully fabricated in the integrated circuit and includes up/down conversion and analog-to-digital conversion (ADC) and digital-to-analog conversion (DAC). Another example of low-cost 60 GHz phased arrays with highly integrated Si-CMOS integrated circuit is given in [11]. Their approach is to invert the package and use antennas on the backside of the package, which is similar to the approach described in [12] and is illustrated in Fig. 2.6. Note how the package is flipped with solder ball and interconnects to the motherboard. The antenna elements that form the phased array create a radiation pattern with broadside radiation that is perpendicular to the motherboard printed circuit board (PCB). A common in these approaches is the use of Si-CMOS for the integrated circuit chip set and low-cost laminate type packaging.

2.3.4 Millimeter-Wave Point-to-Point Systems

Back haul wireless systems are used to carry data traffic from user end points or points where user traffic is aggregated to locations where it can be trunk connected. Increasingly, these systems are operating at millimeter-wave frequencies. Interesting

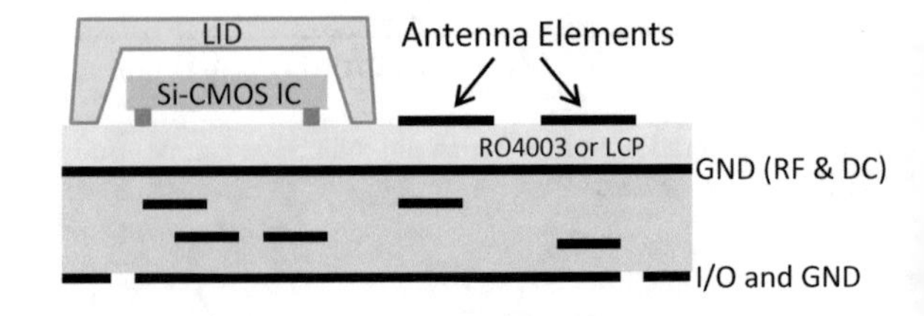

Fig. 2.5 Illustration of how laminates have been proposed for 60 GHz phased arrays on the 802.11ad standard

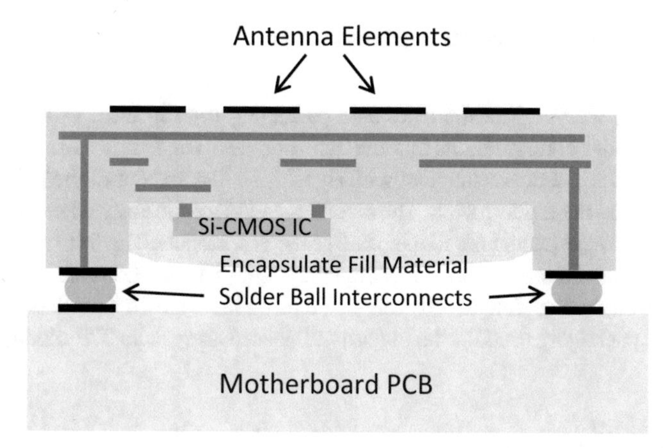

Fig. 2.6 Inverting the IC package allows phased array antennas to be integrated easily into the package

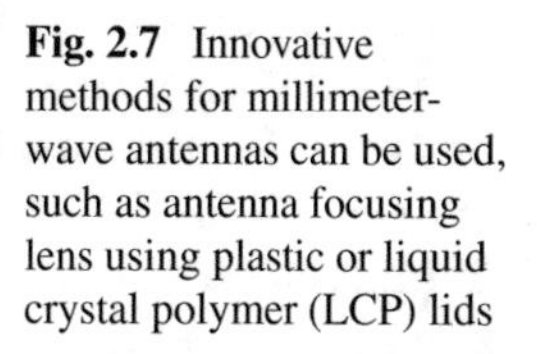
Fig. 2.7 Innovative methods for millimeter-wave antennas can be used, such as antenna focusing lens using plastic or liquid crystal polymer (LCP) lids

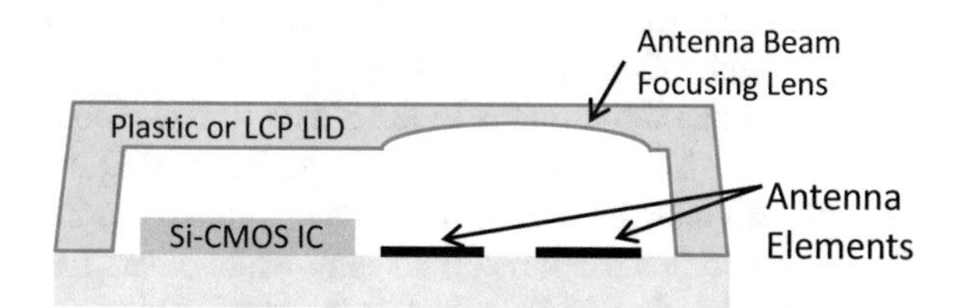

approaches have been developed to achieve low-cost packaging of phased arrays and transmit receive modules. One technique is described in [13] for a 60 GHz backhaul system. It uses a lens in a dielectric lid over an array of patch element as illustrated in Fig. 2.7.

The driving need for high-speed millimeter-wave backhaul networks is that wireless networks traffic is projected to increase 5000 fold over the next 15 years [14]. This means new methods must be developed to handle the increased data. The 60 GHz band is particularly attractive since it is unlicensed, has wide bandwidth of 57–64 GHz (in USA) [15], and has RF max power specified in EIRP. A goal in recent FCC rule changes in the USA is to "allow longer communication distances for unlicensed 60 GHz point-to-point systems that operate outdoors and thereby extend the ability of such systems to provide broadband service [15]." This need has created opportunities for phased arrays and transmit receive modules for millimeter-wave point-to-point systems.

One key advantage that phased arrays provide is their ability to provide functionality required for self-organizing networks (SON) [16]. This is because the ability to electronically scan the antenna beam eases network planning and link installation since the network antenna links can be remotely modified after installation and on-the-fly as the network evolves. A particular attraction is installation and alignment costs will be lower. For these reasons, phased arrays with T/R modules provide capability essential to meeting future point-to-point data communication needs.

2.4 T/R Modules in Phased Array Radar

In this section, we consider the electronic packaging used in three different types of phased array radar configurations that use T/R modules. The first is called a brick array and uses modules that are rectangular with signal flow into and out of the T/R module in the same plane as the module itself. The second is called the tile array where the input and output signals are perpendicular to the plane of the T/R module. The third is a panel array where the notion of a module no longer applies, but the T/R functions are integrated with the antennas onto a flat plate usually fabricated using laminates. These three approaches illustrate the packaging used for the majority of phased array radar T/R modules.

2.4.1 Brick Array

An example of a brick T/R module is shown in Fig. 2.8. There are a few variations on the packaging methods used for it, but basically, it contains a ceramic or laminate substrate with radar signal flow that is essentially restricted in the x- and y-axes. There are short radar signal paths in the z-axis to make contact with buried RF stripline. However, the brick module can be considered a 2D module. Another distinguishing factor is that most brick modules contain a single channel of T/R function per module. Some approaches use a metal housing while others use the ceramic substrate as the housing with a ring frame, which is the case illustrated in Fig. 2.8.

2.4.2 Tile Array

Tile array T/R modules are packaged using ceramic substrates that are stacked on top of each other to form a 3D stack with radar signals flowing in all three dimensions. Each module contains four channels of T/R functions, which is an important distinction. The interconnects between substrates are achieved using solderless interconnects [17].

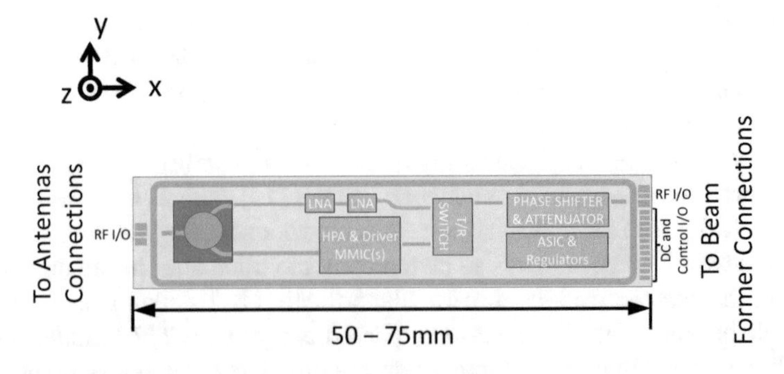

Fig. 2.8 Brick style T/R module

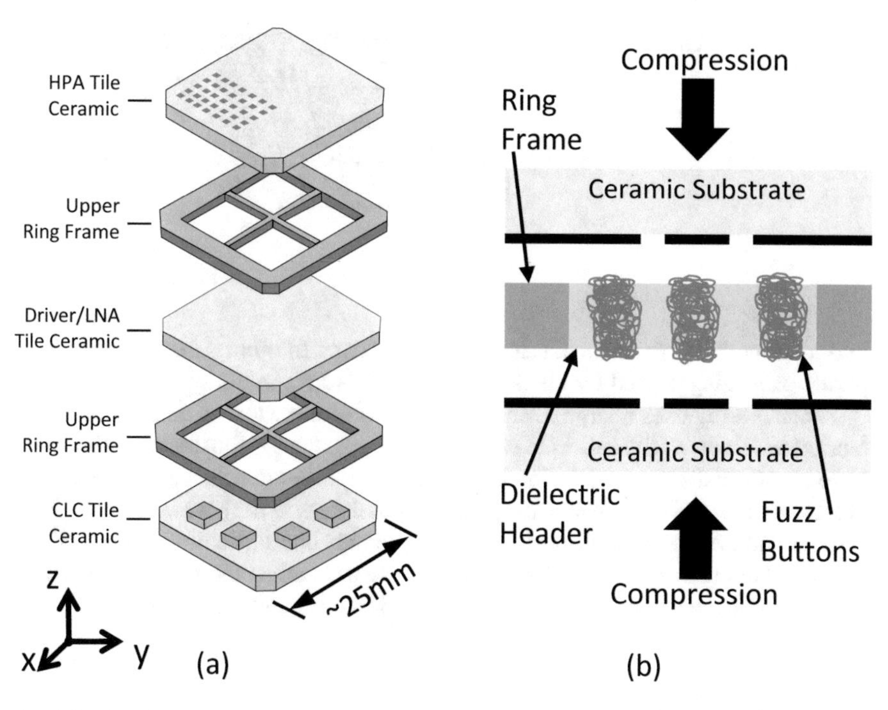

Fig. 2.9 The tile array consists of (**a**) stacked substrates and ring frames to form a 3D stack of substrates and (**b**) solderless interconnects

The configuration is illustrated in Fig. 2.9a. In this design, there are three substrates, each with a particular function for the system. The substrates are separated by aluminum ring frames that also serve the function capturing the solderless interconnects. The solderless interconnects can be formed using several different methods. One approach is to use an elastomeric connector [18, 19]. The elastomeric connector has a multitude of small wire conductors embedded in the elastomer. When it is compressed, the conductive wires make electrical contact between the substrates. Another method to implement the solderless interconnect is to use fuzz buttons [20, 21]. A simplified cross section is illustrated in Fig. 2.9b. Fuzz buttons are wires that are stuffed into a cylindrical shape and placed into a dielectric header. The header is captured by the aluminum ring frame. When the substrates are compressed, the fuzz buttons make contact to the metal conductors on the ceramic substrate. The fuzz button interconnects can be used for bias, control signals, and RF interconnects. The tile array packaging approach is one method to achieve a 3D module that is more compact and lower weight than brick modules.

2.4.3 *Panel Array*

The idea of a separate T/R module does not apply to the packaging of a panel array. This is because the T/R functions are integrated with the antenna and, in most cases, the hermetic module is eliminated. In some cases, the T/R functions are packaged

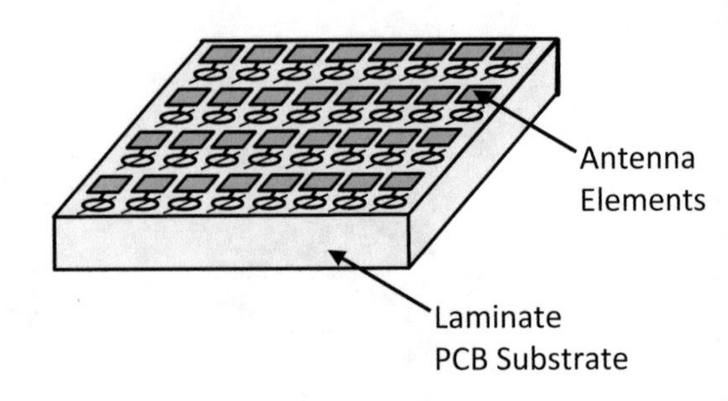

Fig. 2.10 Illustration of a panel array

into a hermetic IC package that is nearly chip scale. In other cases, non-hermetic plastic packaging is used for all of the integrated circuits in the array.

A panel array uses a large format laminate substrate with the antenna elements fabricated as part of the substrate itself. This arrangement is illustrated in Fig. 2.10. Note how each antenna patch element has a phase shifter to achieve beam steering. Most implementations of panel arrays use air cooling as in [22], which also use both vertical and horizontal polarization at each radiator and flip chip SiGe and GaAs integrated circuits. The use of low-cost laminate packaging and silicon-based integrated circuits means that panel arrays can achieve very low cost goals.

2.5 Thermal Packaging Challenges

One of the packaging challenges for T/R modules is the heat dissipated by the high-power amplifiers. The challenge is twofold. First, the amount of heat dissipated can be significant. Consider, for instance, an array with 256 elements that is used in an S-band radar system and each high-power amplifier generating 100 W of output power at 40% efficiency. The dissipated power is given by

$$P_{\text{diss}} = \left(P_{dc} + P_{\text{in}}\right) - P_{\text{out}} = \frac{P_{\text{out}} - P_{\text{in}}}{\eta} + P_{\text{in}} - P_{\text{out}} \tag{2.1}$$

where:

P_{diss} = heat dissipated (W).

P_{in} = RF input power to the high-power amplifier (W).

P_{out} = RF power out of the high-power amplifier (W).

η = efficiency of the high-power amplifier = $(P_{\text{out}} - P_{\text{in}})/P_{\text{dc}}$ (%).

P_{dc} = direct current power required by the high-power amplifier (W).

If we assume that the input power to each amplifier in the array is 1 W, then using (1), we find that the dissipated power is

$$P_{\text{diss}} = \frac{75\text{W} - 1\text{W}}{0.5} + 1\text{W} - 75\text{W} \sim 75\text{W}$$

This means that a 256-element phased array dissipates nearly 19 kW of thermal energy. Of course, this assumes that the amplifiers will be operated at 100% duty cycle, which is not the case for most radar systems. Even if the amplifiers are operated at 25% duty cycle, this is still over 4.7 kW of dissipated power.

As if this problem was not significant enough, the situation is much actually worse when the heat flux is considered for a typical high-power amplifier. Heat flux is the heat density, which can be written as

$$q'' = \frac{Q}{A} \tag{2.2}$$

where:

Q = heat dissipated (W).

A = area of the heat source (cm^2).

An S-band GaN MMIC amplifier is shown in Fig. 2.11. The output stage of the amplifier is where most of the heat is dissipated, the size of the output amplifier field effect transistors is approximately 0.55 mm × 4.42 mm, and the transistor dissipates approximately 75 W of heat. Therefore, the heat flux at the MMIC transistors is

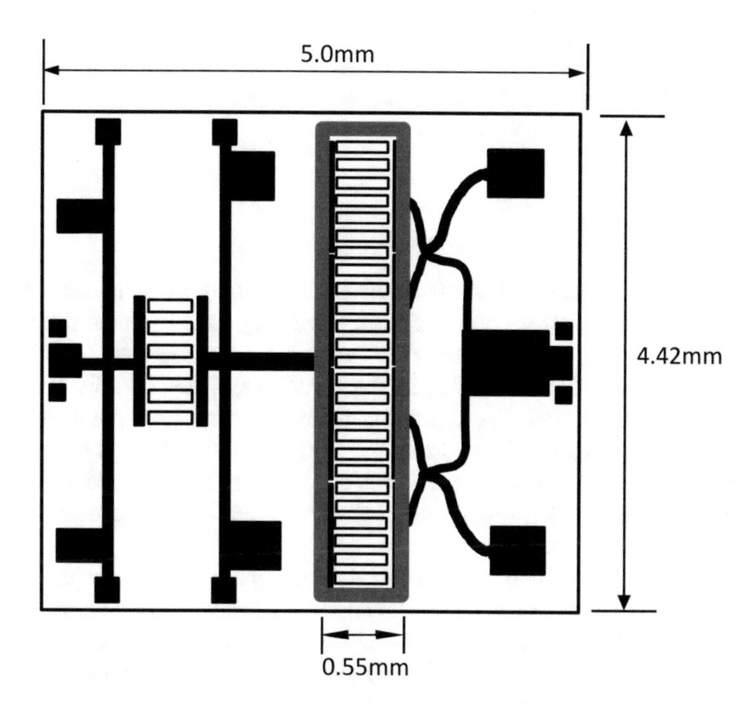

Fig. 2.11 Image of a GaN high-power amplifier showing the heat flux

$$q'' = \frac{Q}{A} = \frac{75\text{W}}{0.42cm \cdot 0.055cm} = 3247\text{W} / cm^2$$

Again, this assumes 100% transmit duty cycle. Most radar systems operate at transmit duty cycles lesser than 100%, which can greatly reduce the average heat flux. If the amplifier in this example is operated at 5% duty cycle , then the average heat flux at the integrated circuit transistors will be approximately 160 W/cm^2. Nevertheless, this level of heat flux requires careful thermal design.

2.6 Wafer Level T/R Module Packaging

For many applications, the lowest cost and most efficient packaging are achieved at the wafer level. This can be done for millimeter-wave T/R modules and phased arrays since the antenna is small enough to be integrated on the semiconductor wafer. At 60 GHz, for instance, a half wavelength is approximately 2.5 mm, which is small enough so that an array of elements can be integrated directly on the silicon wafer. A 64-element phased array was developed at 60 GHz, which occupied a full 2.2 × 2.2 cm^2 reticle in SiGe [23]. The antenna measurements showed beam steering of ±55° in both E- and H-planes with an EIRP of approximately 38 dBm. This method of fabricating 60 GHz phased arrays and T/R modules offers performance and levels that are difficult to achieve with other technologies.

2.7 Conclusions

This chapter described some of the electronic packaging methods used for T/R modules for both commercial applications and military radar. Commercial systems that can benefit from phased arrays and T/R modules were discussed. This includes cellular base stations, Wi-Fi-enabled indoor location systems, 60 GHz Wi-Fi, and millimeter-wave back haul. Packaging examples used in those applications were presented. Three types of phased array radar T/R module packaging approaches were also described. The brick module, tile array, and panel array module packaging were described. Finally, wafer level packaging of millimeter-wave phased array T/R modules was discussed.

Future T/R modules will have reduced functionality in the analog domain. Instead, as digital sampling technology increases in frequency, digital beam forming will become a reality. In these types of systems the packaging of the T/R modules will be much simpler and lower cost with the complexity being transferred to the high-data rate information being generated at each antenna element.

References

1. Alidio, R., W.Y. Lee, A. Gummalla, and M. Achour. 2010. A novel broadband power amplifier architecture for high efficiency and high linearity applications. In *IEEE International Microwave Symposium*, 1064–1067. Anaheim, CA.
2. Caverly, R.H., and M.J. Quinn. 1999. Time domain modeling of PIN control and limiter diodes. In *IEEE International Microwave Symposium*, 719–722. Anaheim, CA.
3. Sturdivant, R., J. F. Brown, and C. Turner. Location determination system and method using array elements for location tracking. US Patent Application 0181867A1, issued July 18, 2013.
4. Mickhayluck, Y.P., A.A. Savochkin, A.A. Schekaturin, and V.M. Iskiv. 2007. Wi-Fi networks user's terminal phased antenna array. In *International Conference on Antenna Theory and Techniques*, 446–448. Ukraine.
5. Wen, F., and C. Liang. 2015. Fine-grained indoor localization using single access point with multiple antennas. *IEEE Sensors Journal* 15(3): 1538–1544.
6. Attanayake, G., and Y. Rong. 2012. RSS-based indoor positioning accuracy improvement using antenna array in WLAN environments. *International Conference on Indoor Positioning and Indoor Navigation.*
7. Pinel, S., S. Sarkar, P. Sen, B. Perumana, D. Yeh, D. Dawn, and J. Laskar. 2008. A 90 nm CMOS 60GHz radio. In *IEEE International Solid-State Circuits Conference*, 130–131 . San Francisco, CA.601
8. Laskar, J., S. Pinel, D. Dawn, S. Sarkar, P. Sen, B. Perunama, D. Yeh, and F. Barale. 2009. 60GHz entertainment connectivity solution. In *IEEE International Conference on Ultra-Wideband*, 17–21. Vancouver, BC.
9. Lee, J., Y. Huang, Y. Chen, H. Lu, and C. Chang. 2009. A low-power fully integrated 60GHz transceiver system with OOK modulation and on-board antenna. In *IEEE International Solid-State Circuits Conference*, 316–317. San Francisco, CA.
10. Laskar, J., S. Pinel, D. Dawn, S. Sakar, P. Sen, B. Perunama, and D. Yeh. 2007. FR-4 and CMOS: enabling technologies for consumer volume millimeterwave applications. In *IEEE International Electron Devices Meeting*, 981–984. Washington, DC.
11. Brebels, S., K. Khalaf, G. Mangraviti, K. Vaesen, et al. 2016. 60 GHz CMOS TX/RX chipset on organic packages with integrated phased-array antennas. In *10th European Conference on Antennas and Propagation*, 1–5. Davos, Switzerland.
12. Bonthron, A. Method and apparatus for automotive radar sensor. US Patent Application 0225481A1, issued Oct. 13, 2005.
13. Pettus, M.G., and J.R.A. Bardeen. Integrated antenna and chip packaging and method of manufacturing thereof. US Patent Number 7,768,457B2, issued Aug. 3, 2010.
14. Evolutionary and disruptive visions towards ultra high capacity networks. *International Wireless Industry Consortium*, April 2014.
15. Revision of Part 15 of the Commission's rules regarding operation in the 57-64GHz band. Federal Communications Commission Report and Order, Released Aug. 9, 2013.
16. Verma, L., M. Fakharzadeh, and S. Choi. 2015. Backhaul need for speed: 60GHz is the solution. *IEEE Wireless Communications* 22(6): 114–121.
17. Hauhe, M.S., and J.J. Wooldridge. 1997. High density packaging of X-band active array modules. *IEEE Transactions on Components, Packaging, and Manufacturing—Part B* 20(3): 279–291.
18. Sturdivant, R., C. Quan, and B. Young. 1994. Using the matrix metal-on-elastomer connector at microwave frequencies. In *27th International Symposium on Microelectronics*. Boston, MA.
19. Ivanov, S.A., and V. Peshlov. Vertical transitions with elastomeric connectors. International Conference on Microwave, Radar, and Wireless Communications, 2002, 472–475.
20. Sturdivant, R., C. Ly, J. Benson, and M. Hauhe. 1997. Design and performance of a high density 3D microwave module. In *IEEE International Microwave Symposium*, 501–504. Denver, CO.

21. Sturdivant, R., and C. Quan. Microwave vertical interconnect through circuit with compressible connector, US Patent 5,552,752.
22. Puzella, A., and R. Alm. 2008. Air-cooled, active transmit/receive panel array. In *IEEE Radar Conference*, 1–6. Rome, Italy.
23. Zihir, S., O.D. Gurbuz, A. Karroy, S. Raman, and G.M. Rebeiz. 2015. A 60GHz 64-element wafer-scale phased-array with full-reticle design. In *IEEE International Microwave Symposium*, 1–3. Phoenix, AZ.

Chapter 3
3D Transitions and Connections

Rick Sturdivant

3.1 Introduction

Vertical transitions are now very common in many RF and microwave products. This was certainly not the case a few decades ago for at least two reasons. First, the simulation tools required to design vertical transitions were not commonly available. Until the early 1990s, three-dimensional (3D) electromagnetic analysis using numerical methods was, in general, restricted to university settings and some research labs. However, by the mid 1990s, 3D electromagnetic simulation and the required computing power were widely available. Second, the fabrication methods and materials were not available or not known to be useful for RF and microwave products. For instance, low temperature co-fired ceramic (LTCC) became widely used after the early/mid 1990s. LTCC is important for vertical transitions since it allows for both blind and buried vias which enables optimized design and layout. In the early 1990s, designers began to experiment with other materials such as high temperature co-fired ceramic (HTCC) in aluminum nitride and alumina to determine their usefulness for RF and microwave applications [1]. Because of widely available 3D electromagnetic simulators and fabrication methods, vertical transitions are used in a variety of RF, microwave, and millimeter-wave products.

Traditionally, RF and microwave signals on transmission lines were routed on planar boards and the signals did not transition out of the x–y plane except for very small z-axis travel for interconnect to integrated circuits. A vertical transition is formed when the signal is transitioned from a transmission line in one plane to a transmission line in another plane. A very common vertical transition is from microstrip to buried stripline which is illustrated in Fig. 3.1. Note how the top transmission line is connected to the buried transmission line using conductive vias in the substrate material.

R. Sturdivant (✉)
Azusa Pacific University, Azusa, California, USA
e-mail: ricksturdivant@gmail.com

K. Kuang, R. Sturdivant (eds.), *RF and Microwave Microelectronics Packaging II*,
DOI 10.1007/978-3-319-51697-4_3

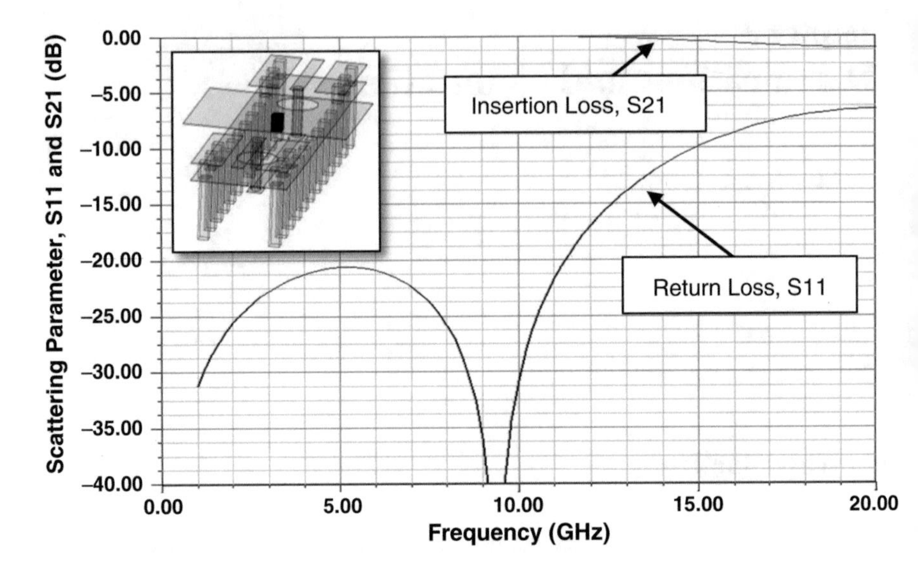

Fig. 3.1 Example of one type of 3D transition using microstrip and buried stripline with the simulated return loss and insertion loss in dB

In this chapter, we will consider four types of 3D transitions. Each transition type has its own challenges, but each follows the design principles of keeping the transition as compact as possible and using transmission lines as the transition element. The transitions considered here are:

1. Vertical Transitions Between Planar Transmission Lines: We will consider the microstrip to stripline transition in co-fired ceramic substrates.
2. Vertical Transitions Using Stacked Die and Through Silicon Vias: In this approach, integrated circuits are stacked on top of each other and interconnects are made between them.
3. 3D Transitions Using Connectors: Transition connectors transfer the RF signal from one layer to another layer in the packaging. The fuzz button connector and elastomeric connectors were mentioned in the previous chapter. In this chapter, the SMP connector is described.
4. Vertical Transition Using Balls or Bumps: These vertical interconnects are sued with flip chip ICs.

3.2 Vertical Transitions Between Planar Transmission Lines

Transitions between planar transmission lines are very common in RF and microwave products such as mobile phones and satellite systems. This is partly due to the widespread use of circuit boards with planar traces fabricated on surface layers and interior layers. Physically connecting signals from one portion of a circuit board to another normally requires conductor lines to be routed on several different layers. For DC bias lines and low frequency data or control lines, minimal care is required for signal routing. However, as the frequency of operation for the data and control lines increases, the need for transmission lines and carefully designed transitions increases. In this section, we will consider the vertical transition formed by a surface transmission line and a buried transmission line.

3.2.1 Microstrip or Coplanar Waveguide to Stripline

There are a few methods that can be used to transition from microstrip to stripline using a vertical transition. One method uses three-wire lines to make the connection between layers. This approach is illustrated in Fig. 3.2. Note that the microstrip line has topside ground shield on both sides of the transmission line which is often used to reduce coupling between the microstrip and adjacent components. Also notice how the signal line of the microstrip and the stripline are connected by a single via. The topside shielding grounds of the microstrip are connected to both grounds of the stripline transmission line. The three-wire line can connect to stripline in a few different ways and the differences are subtle but important. In one case, the bottom ground for the microstrip line, or another ground layer below it, is used as the topside ground of the stripline. This method is important for cases where the stripline section must be buried low within the substrate to allow for component placement or routing of other signals. This approach was used in [2, 3].

Another approach to implement the three-wire transition between microstrip and stripline uses the ground on the topside of the substrate as the top ground of the stripline and separate lower grounds are used as the bottom stripline and microstrip grounds as shown in Fig. 3.3. Note from the figure that the electromagnetic analysis shows the electrical bandwidth is at least 20 GHz. In fact, when careful design methods are used as in [4, 5], it is possible to achieve vertical transitions of this type with electrical bandwidth from 10 MHz to 100 GHz.

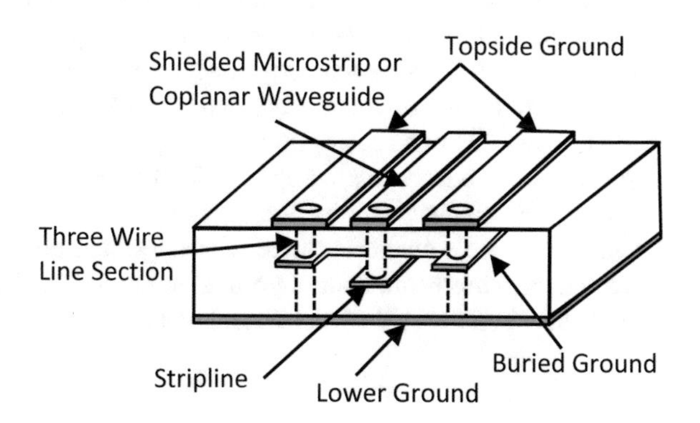

Fig. 3.2 A three-wire line can be used to transition between microstrip (with topside ground shields) or coplanar waveguide and stripline in the case where the stripline is buried lower layers within the multilayer substrate

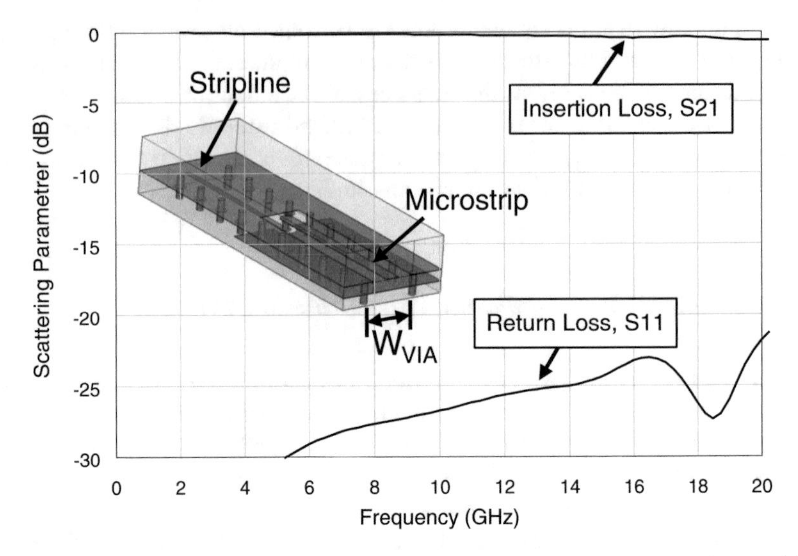

Fig. 3.3 Three-wire line can also be used to transition between shielded microstrip to stripline with a shared topside ground

One of the drawbacks of using three-wire lines as the vertical transmission line is that by themselves, they provide limited isolation. As a result additional vias are often used to minimize the amount of radiation within the substrate. When these additional vias are used for isolation, it is important that their size and placement be chosen carefully to avoid supporting waveguide type modes which may have cutoff frequencies at or below the operating frequency.

Instead of using three-wire lines to form the vertical transition, it is more common for the transition into the buried lines to be realized using a quasi-coaxial transmission line. It is formed using a ring of vias around the center signal conductor and the via ring exists between buried metal ground planes. Figure 3.4a shows a cross-section of an ideal coaxial line with a center conductor and solid outer conductor. For comparison, Fig. 3.4b shows the cross-section of a three-wire line. Figure 3.4c shows the quasi-coaxial line where the center conductor is surrounded by a ring of vias. The number and location of the vias depends upon the fabrication rules of the multilayer process being used. A simplified model of the transition is shown in Fig. 3.4e. Note that the inductance in the model is meant to model the inductive effect of the via and the capacitance is meant to model the capacitive effects of the 90° bond from the microstrip to via. A more complex model for a vertical transition from microstrip to stripline can be found in [6].

Other workers have studied the effect of various transitions types using only a single via on the signal line, three-wire lines, and quasi-coaxial line with several different cases for the number of ground vias used [7]. Their measured data showed that for the single via case its electrical bandwidth was approximately 64 GHz and

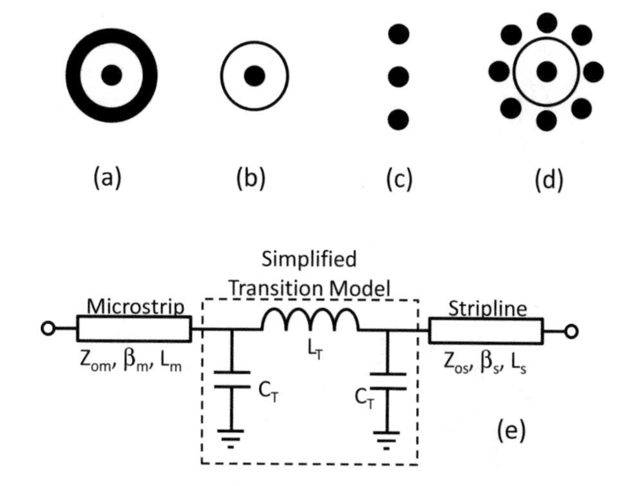

Fig. 3.4 Transition transmission lines may include (**a**) ideal coax, (**b**) quasi-coax formed by a single via in an opening in a ground plane, (**c**) three-wire transmission line, (**d**) quasi-coax formed using a via through a ground plane with a ring of vias connecting ground from layer to layer, and (**e**) a simplified model of the vertical transition section from microstrip to stripline

for the quasi-coaxial transition with a dense pack of 12 ground vias the bandwidth was approximately 90.5 GHz.

A variation on this approach was taken in [8] except that the grounds for the coaxial section were fabricated using curved slots so that the ideal coax line was better approximated. Their measured results showed the transition functioned to approximately 40 GHz.

3.2.2 *Top Side Microstrip to Bottom Side Microstrip*

Another closely related variation of this transition is shown in Fig. 3.5. The figure shows a topside microstrip line and a bottom side microstrip line with a three-layer quasi-coaxial transition between them. This type of transition allows signals to be carried through multiple layers of dielectric with excellent signal integrity. The figure solid lines show the transition functions well to approximately 20 GHz and with proper optimization the bandwidth can be extended. One of the limits on the electrical bandwidth is the excitation of higher order modes in the quasi-coax section which must be avoided. The figure also shows that with a simple series capacitive section of transmission line, the transition can be tuned to perform well to over 50 GHz.

Instead of using vias to transition from one microstrip to another in a multilayer structure, it is possible to transition using apertures in the ground plane. This approach was demonstrated in [9, 10]. The ground plane is shared between a top microstrip line and a bottom microstrip line. The common ground plane is opened so that the electromagnetic energy can couple through the slot and onto the opposite microstrip line. Since this approach uses resonant effects in the slot and microstrip line configuration at the transition, it has less electrical bandwidth.

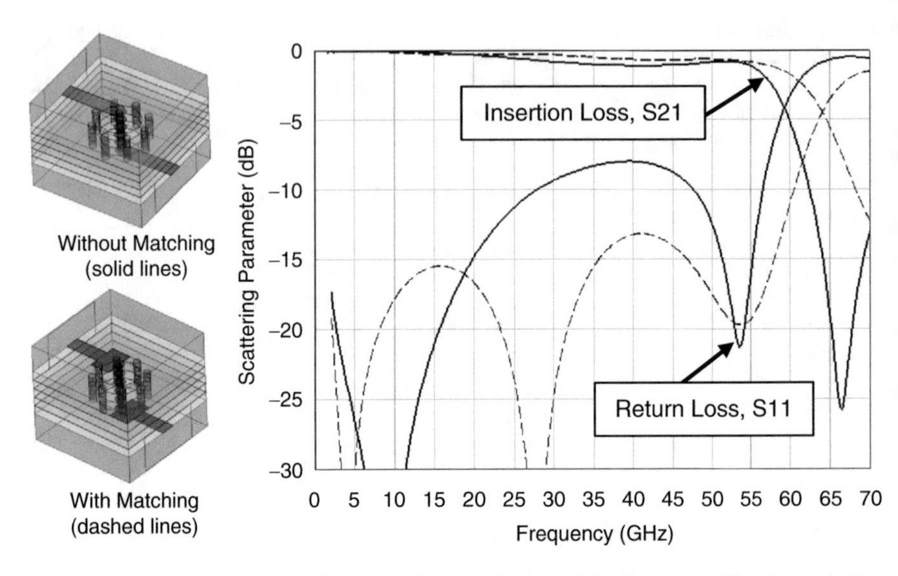

Fig. 3.5 Electromagnetic simulation results for a vertical transition from a topside microstrip line to a bottom side microstrip line in LTCC (ε_r = 5.9) using a three-layer quasi-coax line through (*solid line* is without any matching circuit and *dashed line* is for simple series capacitance matching circuit)

3.2.3 Microstrip to Waveguide Transition

Another transition type connects microstrip to waveguide. This type of transition is important for millimeter-wave systems such as automotive radar, 60 GHz high data rate communication, and E-band back haul systems. The attractiveness of this transition is due to the fact that integrated circuits best interconnected using microstrip (or other planar transmission line) and antenna connections often require waveguide at millimeter-wave frequencies. The concept for this type of transition is illustrated in Fig. 3.6 which has been used for many decades. Note that the probe is normally placed at a quarter wavelength from the waveguide back short. The probe is inserted into a small hole in the broadside wall of the waveguide. The view in the figure is from the narrow wall of the waveguide. MMICs in this illustration are interconnected with wire bonds but other interconnects such as ribbon bonds or flip chip can be used. More recent variations on this transition are integrated with FR-4 [11] and can also use coupled microstrip lines [12].

3.3 Through Silicon Via 3D Transitions

Through silicon via (TSV) interconnects are attractive because the size of the vias can be extremely small and the pitch between vias can be very tight. For instance, 25 μm diameter vias have been fabricated in 100 μm silicon with 75–100 μm pitch between

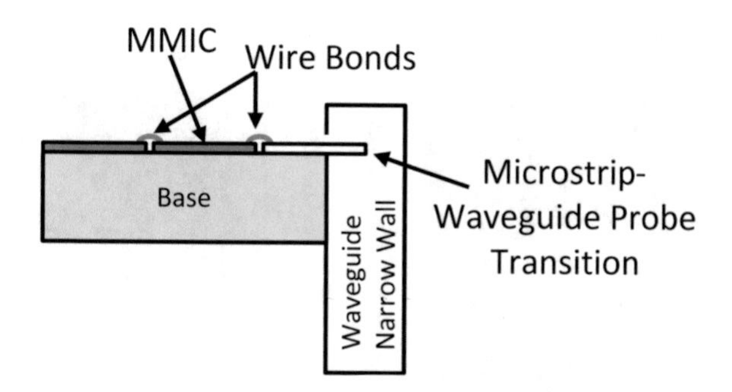

Fig. 3.6 Example transition from microstrip to waveguide through the waveguide broadside wall

vias. The applications for the through silicon technology include system-on-chip where multiple technologies must be combined [13]. Other uses include fan out from tightly spaced integrated circuits and test systems with bandwidths of 40 GHz demonstrated [14].

Because of the small feature size of TSVs, their use can shrink the size electronic equipment such as laptops and mobile phones. However, an even more significant advantage is that they can allow for integration of digital, analog, RF, and microwave circuits using stacked semiconductor die [15, 16]. In this arrangement, active semiconductor die are processed with TSVs and solder bumps are used to connect from one layer to the next.

Because of the size of solder bumps, methods have been investigated to eliminate them for TSV applications. One approach removes the bumps and uses contactless capacitive interconnects between the layers [17]. In one such approach, the signal is passed from one TSV layer to the next using small parallel plate type capacitors for AC signal transfer. The idea is that at high microwave and millimeter-wave, the required series capacitance is small enough to enable capacitive signal transfer between layers.

3.4 Vertical Transitions Using Connectors

Connectors can be used to implement vertical transitions. There are many standard vertical connector types, but a photo of a very common version is shown in Fig. 3.7a. It shows a SubMiniature A (SMA) connector which is solder attached to a circuit board. The bottom side view shows the ground pins and signal pin from the SMA connector soldered to the PCB and the matching circuit near the center pin. Figure 3.7b shows the 3D electromagnetic simulation results and measured data of the connector mounted onto a section of the circuit board. The microstrip line has a short section of matching circuitry which was optimized for 50 Ω impedance match.

A popular type of connector used for vertical connections in RF and microwave circuits is the SMP. One of the attractive features making it useful for connecting between circuit boards is that it can accommodate radial misalignment between the

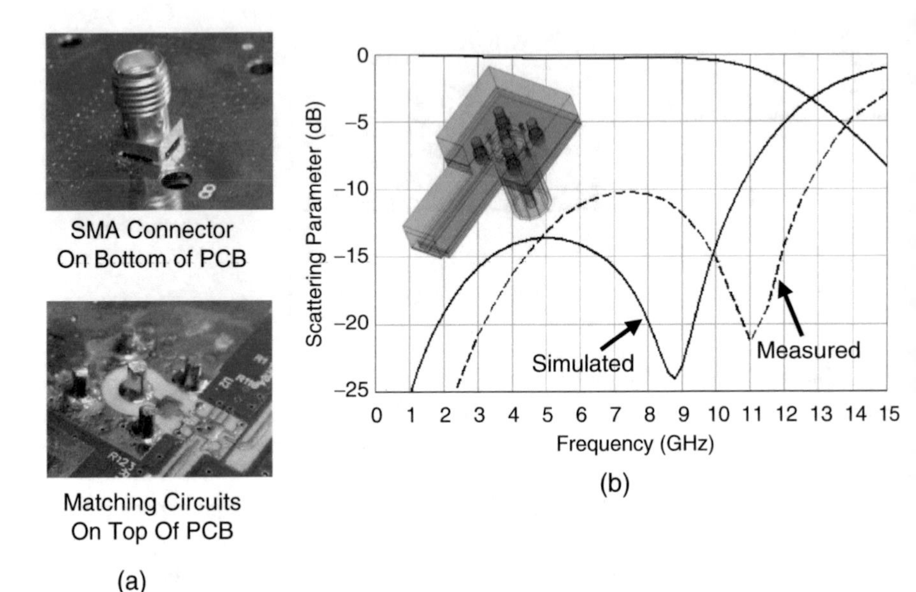

Fig. 3.7 Example of a SMA to PCB transition showing images of the test board, model simulation result in HFSS, and measured data

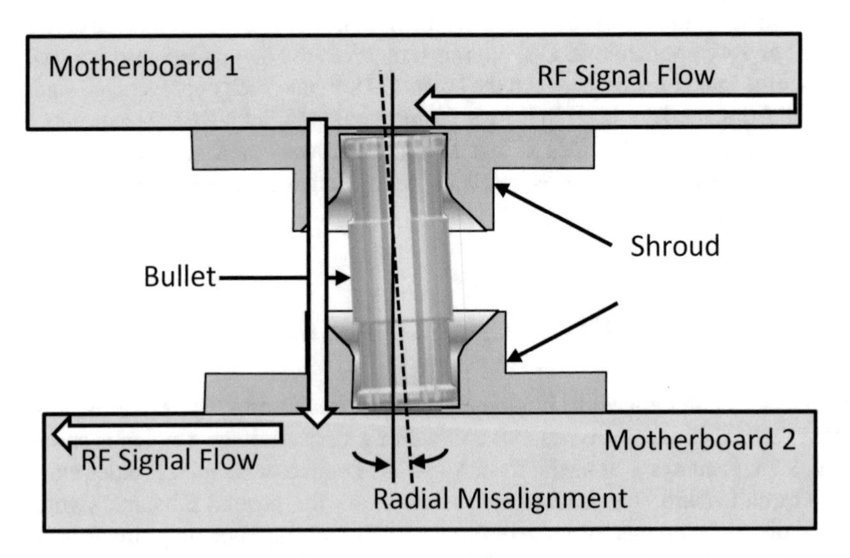

Fig. 3.8 Illustration showing how push on connectors can be used to connected vertically between circuit boards

boards. This is important since mating motherboards can be misaligned due to the tolerances in the board fabrication, SMT component placement, temperature variations, and other manufacturing variability. As a result, a connector that can accommodate these variations and maintain acceptable performance at RF and microwave frequencies is very useful. Figure 3.8 illustrates the SMP connector between two motherboards. The connector shrouds are solder attached to the motherboard and are shown to be misaligned relative to each other. The shrouds include a connector pin that is accepted by the connector bullet. The SMP connector bullet connects between the two shrouds and accommodates the misalignment between them. Note how the signal flow is in the plane of the top motherboard, with a 90° bend into the SMP bullet, and another 90° bend into the plane of the lower motherboard. This type of connector is popular for use from 0.1 to 40 GHz when the appropriate shroud and bullet are chosen.

3.5 Vertical Transition Using Flip Chip

For some RF and microwave applications, flip chip interconnects offer a supplier electrical signal connection. This is due to the fact that the flip chip vertical transition can be designed to function to 100 GHz or more [18–20]. It is common for the connection to be made using solder bumps though other methods are used such as hard bumps and thermosonic gold bumps have also been used [21].

3.6 Conclusions

Vertical transitions and connections are now a common feature of RF and microwave products. Many mobile phones, for instance, use micro miniature surface mount vertical transition connectors to miniature coaxial cables for connection between antennas and amplifiers, switches, and filters. High volume millimeter-wave communications modules use microwave to waveguide transitions so that chip interconnects such as wire bonds and flip chip can be used and waveguide antennas can properly connected. This chapter has introduced several of the more common 3D transitions and connections used in modern RF and microwave products.

References

1. Midford, T.A., J.J. Wooldridge, and R.L. Sturdivant. 1995. The evolution of packages for monolithic microwave and millimeter-wave circuits. *IEEE Transactions Antennas and Propagation* 43(9): 983–991.
2. Sturdivant, R., C. Quan, and J. Wooldridge. 1996. Transitions and interconnects using coplanar waveguide and other three conductor transmission lines. In *IEEE MTT-S International Symposium Digest*, 235–238. San Franscisco, CA.
3. Sturdivant, R. Direct three-wire to stripline connection. US Patent 5,689,216, issued Nov. 18, 1997.
4. Leib, M., M. Mirbach, and W. Menzel. 2010. An ultra-wideband vertical transition from microstrip to stripline in PCB technology. In *Proceedings of the IEEE International Conference on Ultra-Wideband*, 1–4. Nanjing.

5. Deepukumar, D.M., W.E. McKinzie III, B.A. Thrasher, and M.A. Smith. 2013. A 10 MHz to 100 GHz LTCC CPW-to-stripline vertical transition. In *IEEE International Microwave Symposium Digest*, 1–4. Seattle, WA.
6. Sturdivant, R. 2013. *Microwave and Millimeter-Wave Electronic Packaging*, 183–190. Norwood, MA: Artech House.
7. Amaya, R.E., M. Li, K. Hettak, and C.J. Verver. 2010. A broadband 3D vertical microstrip to stripline transition in LTCC using a quasi-coaxial structure for millimeter-wave SOP applications. In *Proceedings of the European Microwave Conference*, 109–112. Paris.
8. Decrossas, E., et al. 2015. High-performance and high-data-rate quasi-coaxial LTCC vertical interconnect transitions for multichip modules and system-on-package applications. *IEEE Transactions on Components, Packaging and Manufacturing Technology* 5(3): 307–313.
9. Tao, Z., J. Zhu, T. Zuo, L. Pan, and Y. Yu. 2016. Broadband microstrip-to-microstrip vertical transition design. *IEEE Microwave and Wireless Components Letters* 26(9): 660–662.
10. Yang, L., L. Zhu, W.W. Choi, and K.W. Tam. 2016. Wideband vertical microstrip-to-microstrip transition designed with cross-coupled microstrip/slotline resonators. In *Proceedings of Asia-Pacific Microwave Conference*, 1–3. Nanjing.
11. Purden, J., D. Zimmerman, and M. Miller. 2012. A new vertical transition for FR-4 based millimeter-wave MCMs. In *IEEE International Microwave Symposium Digest*, 1–4. Montreal.
12. Tong, Z., and A. Stelzer. 2012. A vertical transition between rectangular waveguide and coupled microstrip lines. *IEEE Microwave and Wireless Components Letters* 22(5): 251–255.
13. Chang, K.F., R. Weerasekera, and S.S. Bhattacharya. 2015. Electrical transmission characteristics of vertical transition with through silicon vias (TSVs) in 3D die stack. In *IEEE Electronics Packaging and Technology Conference*, 1–4.
14. Krohnert, K., et al. 2016. Gold TSVs (through silicon vias) for high-frequency III-V semiconductor applications. In *IEEE Electronic Components and Technology Conference*, 82–87. Las Vegas, NV.
15. Williams, C.K., and R.E. Thomas. 2008. Electronic packaging including die with through silicon vias. US Patent Number 7,317,256.
16. Katti, G., M. Stucchi, K.D. Meyer, and W. Dehaene. 2010. Electrical modeling and characterization of through silicon via for three-dimensional ICs. *IEEE Transactions on Electron Devices* 57(1): 256–262.
17. Lee, H., et al. 2016. Signal integrity of bump-less high-speed through silicon via channel for Terabyte/s bandwidth 2.5D IC. In *IEEE Electronic Components and Technology Conference*, 2519–2522. Las Vegas, CA.
18. Jentzsch, A., and W. Heinrich. 2001. Theory and measurement of flip-chip interconnects for frequencies up to 100GHz. *IEEE Transactions on Microwave Theory and Techniques* 49(5): 871–878.
19. Heinrich, W. 2005. The flip-chip approach for millimeter-wave packaging. *IEEE Microwave Magazine* 6(3): 36–45.
20. Chen, L., J. Wood, S. Raman, and N.S. Barker. 2012. Vertical RF transition with mechanical fit for 3-D heterogeneous integration. *IEEE Transactions on Microwave Theory and Techniques* 60(3): 647–654.
21. Felton, L.M. 1994. High yield GaAs flip-chip MMICs lead to low cost T/R modules. *IEEE International Microwave Symposium Digest* 23–27: 1707–1710.

Chapter 4
Electromagnetic Shielding for RF and Microwave Packages

Nozad Karim

4.1 Introduction

Antennas and electromagnetic radiation helped mankind to invent radio, TV, GPS, wireless communication, and many other advanced and convenient technologies that are taken for granted in our daily life. Just for a moment, imagine the world without radiation phenomena, a world without antennas and wireless communication. It would be a world without cellphones, TVs, radios, radar and satellite communication to assist in navigating the sky and oceans, a world in the 1800s. Intentional radiation is necessary to make mobile communication possible; however, unintentional radiation is an obstacle to wireless and wired communications, destructive to electronic systems, and harmful to health.

Electromagnetic radiation has been increasing with every new wired and wireless device introduced to the marketplace. The main sources for the electromagnetic radiation broadcasted from the indoor or outdoor electronic devices are intentional and unintentional waves and noises. Many devices such as cordless and wireless phones, printers, computers, internet of things (IoT)/wearable devices, remote carriage door openers, automotive electronics, base-stations antennas, and other electronics devices emit noise from residencies, workplaces, streets, shopping places, underground subways, and even from airplanes in the sky as illustrated in Fig. 4.1.

Some pundits call the electromagnetic radiations electronic pollution since they are man-made and not normal to natural habitats. Dr. Magda Havas explained that biological effects of radio frequency radiation have been documented and range from cancers to cognitive disorders and sleeping dysfunction among humans, and abnormal behavior, reduced milk yield, miscarriages, and premature deaths among farm

N. Karim (✉)
Amkor Technology, Tempe, Arizona
e-mail: Nozad.Karim@amkor.com

K. Kuang, R. Sturdivant (eds.), *RF and Microwave Microelectronics Packaging II*,
DOI 10.1007/978-3-319-51697-4_4

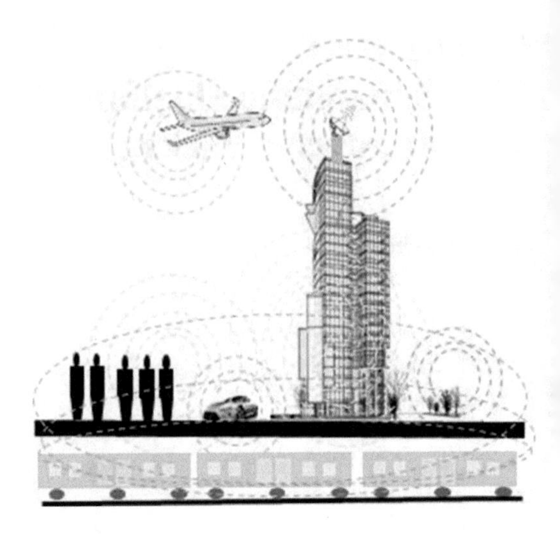

Fig. 4.1 Common sources of radiation emissions

animals [1]. The effects of electromagnetic radiation on human and other biological systems depend both upon the radiation's intensity, duration, and its frequency range.

North American, European, and most Asian electromagnetic compliance commissions' efforts focus on ensuring that electromagnetic compliant products utilizing the latest technologies are safe for consumers and the environment. This involves the entire lifecycle of products in the marketplace [2]. RF, microwave, and digital engineers' task is to eliminate or minimize all unintentional radiations from the electronic devices and components.

Electromagnetic waves propagate through different media, including free space in the radiation mode. Radiation problems can be categorized into two subcategories:

A. Emission
B. Susceptibility

Emission pertains to radiated energy and noise generated by electronic components and systems, imposed on other components, systems and external environment. Susceptibility refers to the sensitivity of electronic components and systems to externally generated electromagnetic noise. This chapter primarily explores solutions to shield RF packages from electromagnetic emissions and minimize the impact of the external noise on the internal electronic circuit elements.

4.2 Electromagnetic Radiation

Today's RF/microwave devices come either in a standalone RF package or in a System-in-Package (SiP). The packages or SiPs are composed of RF, digital and analog devices in addition to passive components such as resistors, capacitors, inductors, SAW filters, crystals, and antennas as well as microelectromechanical systems (MEMS) sensors and more. All the active and non-active components interact with each other in a very condense and small area; in some case, the components in the SiP will be stacked or mounted on top and bottom of the substrate or embedded into the substrate. Advances in SiP and system miniaturization demand new approaches for electromagnetic performance enhancements to be implemented as a part of RF package design flow.

The analysis of radiated emissions is more complex and difficult to understand than conducted emissions; it is both an art and science. The science part comes from a deep understanding of the electromagnetic theory and mathematical relations among all the elements; the art part comes from many years of design and field experience in various electromagnetic design approaches and methods. Impedance mismatch, high level of signal return loss, improper VSWR and standing wave levels are main sources for unintentional radiation. It is essential for RF/microwave engineers to understand digital and analog noise and their impact on RF circuits. For example, while RF input sensitivity might be within the microvolt (μV) level or sub-μV level, digital noise in the system can be in millivolt (mV) level. This means that digital noise can exceed RF noise by a factor of a 100–1000 times. As a result, digital noise coupling can have severe consequences on an RF circuit's performance. RF circuits need to be well protected from digital circuits' conductive and radiative noise.

Digital circuits generate periodic and non-periodic noise that has a significant impact on the RF circuits. The clock frequency provides synchronization to the system and most systems use more than one clock with unrelated harmonic frequency sequences. The clock frequencies carry high current at the fundamental frequency with strong harmonic frequencies harmful to the RF circuitry that operates in proximity of those harmonic frequencies and become one of the major sources for unintentional radiation.

Additional sources of noise in SiP designs are power supplies, DC/DC converters, and power management integrated circuits (PMICs). A power supply's switching frequencies, typically around 50 kHz–10 MHz with high-current amplitude, are the major source of low-frequency radiation. Switching power supply harmonics can be the dominant source of all undesirable and harmful radiation impacts on MEMS and sensors within SiP products.

The other source of electromagnetic radiation is non-periodical signals. These are random noise as a product of certain arithmetic and logic processes that do not repeat at regular intervals. For example, typical computer programs generate signals which are, for practical purpose, random [3]. Other random noises, such as white noise, affect all the frequency components of a signal equally with Gaussian random variables. In contrast, pink noise affects different frequency components differently.

4.2.1 The Source of Radiation

Changes in voltage and current forms are the source of electromagnetic waves. Moving charges in a free space or a specific media initiates electric and magnetic fields in the form of electrical and magnetic waves and transfers power from point A (transmitter) to point B (receiver). Electromagnetic waves move at the speed of

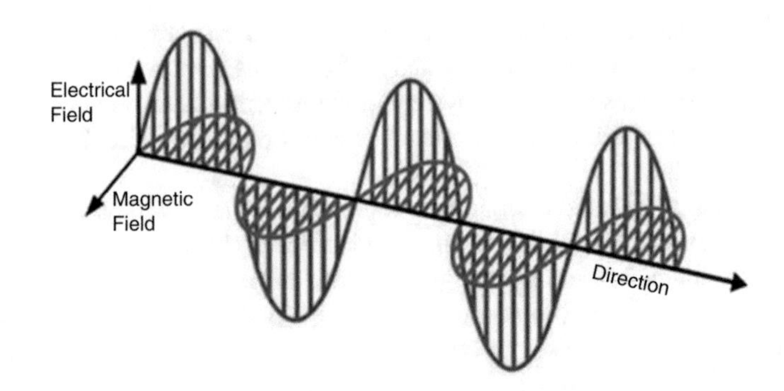

Fig. 4.2 Electromagnetic and magnetic field waves

light in a free space and slow down by the inverse of the square root of the propagation media's dielectric constant. Electromagnetic waves consist of alternating electric and magnetic fields, perpendicular to each other as shown in Fig. 4.2, following the right hand rule. Assume the right hand is in hand shake position, with the thumb pointed upward in a direction to the magnetic force, the fingers pointed forward representing moving charge, i.e., the direction of the velocity. The magnetic field is directly perpendicular to the palm, and it can be represented by curling the fingers 90°to the left. Both fields are interdependent: moving charges generate a magnetic field and a changing magnetic field generates an electric field. Magnetic field has no impact on charges are in static/still mode.

The charges moving on an SiP substrate in a loop form as traveling (to) and returning (from) paths between the source and the load. The return path can be either a trace or a plane; however, a good RF design practice requires a solid plan for the return path. The inductance of the return path is directly proportional to the distance between the driven trace and the return path through wide trace or power and ground planes. One can calculate the voltage drop across the return path of the SiP substrate power and ground planes. The voltage potential change in the low impedance plane will act as a signal source ready to radiate if it is attached to any stub, test point, or wire with good matching characteristic to the source impedance at a specific frequency bandwidth. The wire or the stub will act as a dipole antenna if its active length approaches half the wavelength of the operating frequency, the harmonics of the signal and clock frequencies, or any other non-periodical signals.

Power or ground plans in an SiP can be a solid plane or non-solid plane. The actual power and ground planes can be divided into many islands with different voltage potentials, in addition to many cuts and discontinuities on the plane due to signal vias with keep-out zones passing through both ground and power planes or islands. Models proposed by many researchers predict that emissions will rise dramatically if the return plane interrupts the signal return path with a cut or a slit [4].

4.2.2 How an Antenna Radiates Electric and Magnetic Fields

Frequency, wavelength, and velocity are distinguishing characteristics of electromagnetic waves. A wave travels at constant speed at 299,792,458 m/s in a free space regardless of its frequency and wavelength. The speed of the wave degrades moving from free space to different media as mentioned previously.

The relation between frequency (f) and the wavelength (λ) of a sinusoidal waveform traveling at constant speed (V) is given by:

$$\lambda = \nu / f \tag{4.1}$$

Hofhiens [5] simplified the wave propagation concept with a typical parallel-plate capacitor illustration. Figure 4.3 demonstrates the concept attaching a DC source to a capacitor similar to the decoupling capacitors in power lines. One of the plates continuously holds a positive charge, and the other plate continuously holds a negative charge. A voltage applied to both plates produces an electric field between the plates.

By replacing the DC source with AC source (alternate signal) and changing the ordination of the capacitor plates as shown in Fig. 4.4, the charges on both plates change to represent two opposite sequences of positive and negative charges. Figure 4.4 also depicts the significant changes in the electrical field direction that alternates with AC signal voltage-level changes. Assume the length of the plates equal the half wavelength of a specific frequency to represent a dipole antenna with the supplied AC voltage connected to the middle of the antenna. As the signal voltage polarity alternates at the frequency specified by the length of the dipole antenna, an alternating electric field originates from the antenna, and radiates outward in loops.

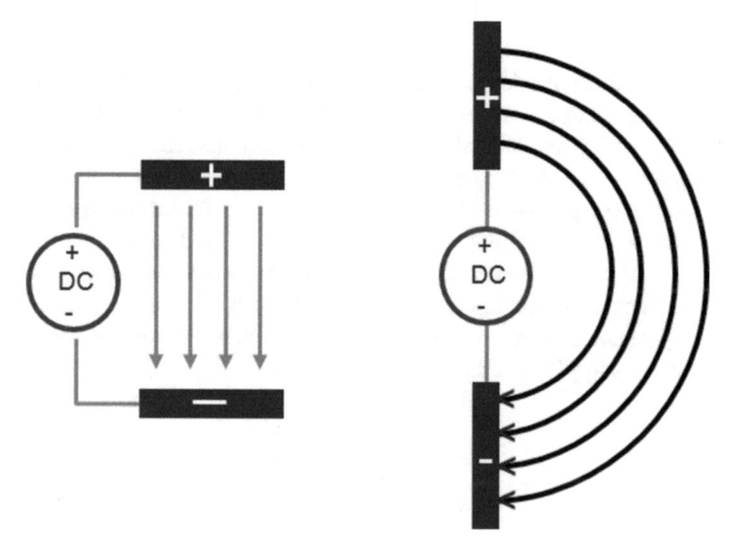

Fig. 4.3 Parallel plate model with a DC source

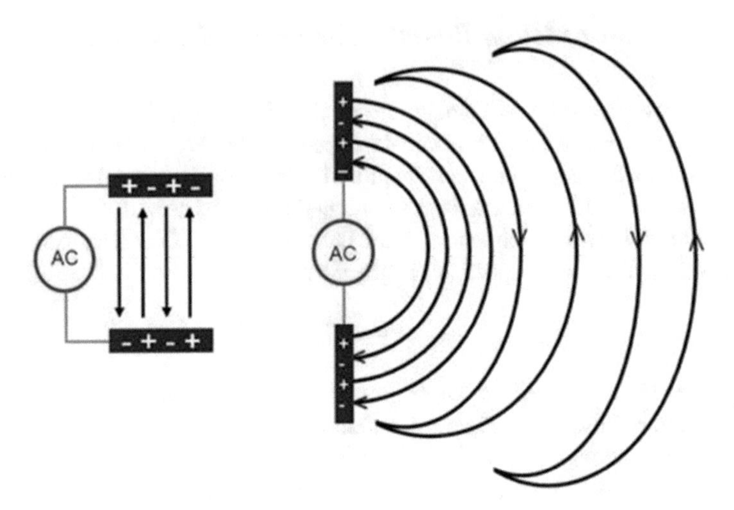

Fig. 4.4 Parallel plate model with an AC source

Also, the current direction alternates leading to alternating direction of the magnetic field loops to travel outward and perpendicular to the antenna. The loop shape makes it possible for alternating electrical fields to propagate through a medium. The direction of the signal trace on the SiP substrate and vertical interconnects through wire bonds, bumps, or vias dictate the polarization for the electromagnetic wave. Generally speaking, SiP devices may have multiple radiation sources with different polarizations and radiation directivities and intensities.

4.2.3 Theoretical View of Radiation

Accelerated charged particles produce electromagnetic waves and can subsequently interact with any charged particles in their field. Maxwell discovered that a varying electric field is always associated with a magnetic field that changes over time. Both electromagnetic wave fields propagate out into space and never again affect the source, except some portion of the wave reflects back to the source from a shielding wall. Maxwell's Equations defined the electromagnetic field with relation to voltage potential and current. However, the equations are not easily understood since they are a set of partial differential equations that consist of integral and differential equations. Glen Dash expressed Maxwell's Equations in a computational form that allows computers to perform the calculations [6]:

$$E = -\left(\nabla V + \frac{\partial A}{\partial t}\right)$$

$$B = \nabla \times A$$

$$V = \frac{1}{4\pi \in} \sum_{n=0}^{n=N} \frac{\rho_n}{r_n} v_n$$

$$A = \frac{\mu}{4\pi} \sum_{n=0}^{n=N} \frac{J_n}{\mathrm{r}_n} I_n a_n$$

Volume:

$$V_n = I_n a_n$$

where:

∇ = Gradient
E = Electric field in V/m
B = Magnetic flux density in tesla (T) or weber (Wb)/m^2, $B = \mu H$
μ = Permeability
H = Magnetic field in Amps/m
V = Voltage in volts
A = The "vector potential" in volt-second per meter V s m^{-1}
p_{n} = Charge density in Coulombs/m3 of a particular charge element, n
r_n = Distance from a given charge or current element, n to the location of interest
V_n = Volume of a particular charge element in m^3, n
l_n = Length of a particular current element in m, n
a_n = Area of a particular current element in m^2, n
J_n = Total current density (both conductive and displacement) in amps/m^2 of a particular current element, n

The vector potential can be calculated by knowing the current density and volume of particular current element. The gradient of the vector potentials leads to the magnetic flux density. Voltage can be calculated from knowing charge density and volume of particular charge element. Adding the gradient of voltage to the derivative of the vector potential leads to the electrical field.

4.2.4 Electromagnetic Simulation and Computational Method

RF and microwave SiP and similar packages must be designed carefully to prevent costly radiation and susceptibility problems in targeted applications, before building prototypes or releasing the device for production. The best way to identify a radiation source and potential problem in the system is to use proven and reliable simulation tools with the appropriate computational algorithms. In modern systems, computer-aided analysis and optimization have replaced the design process of iterative experimental modification of the initial design [7]. However, a high dependency on electromagnetic simulation tools in pre-design, design, and post-design stages of shielding structures may cause correlation problems with the actual device performance.

Every simulation tool or modeling method has its own strengths and weaknesses. Over dependency on simulation without adequate experience in shielding and antenna design methods as well as a deep understanding of simulation techniques might lead to poor shielding solutions with severe consequences. There are different methods to solve Maxwell's electromagnetic equations to predict the wave propagation. Several different simulation or modeling methods were found in current electromagnetic interference (EMI) simulation software packages. The most common of these are:

- The Finite Element Method (FEM)
- The Method of Moments (MoM)
- Finite-Difference Time-Domain (FDTD)
- Frequency-Domain Finite-Differences (FDFD)

Each of these methods has its recommended area of application for maximum efficiency, as well as areas of deficiency where analysis becomes inaccurate and time consuming. Many commercial and free software packages combine desirable techniques into a multiple stage modeling to optimize simulation time and simulation accuracy. The potential causes of the disparity between simulated and measured values are from imperfections of the shielded module structure, unknown material properties at the different frequencies and the test setup that are difficult to take into account to achieve accurate correlation between simulated and measured data.

4.2.4.1 The Finite Element Method

FEM, also referred to as finite element analysis (FEA), is a computational numerical technique used to achieve approximate solutions of boundary value problems for partial differential equations. The technique subdivides a large problem into smaller finite elements using mesh generation techniques. FEM solves simpler equations representing the finite elements then assembles them into a larger system of equations that models the entire structure. It accurately represents complex geometry with the presence of different material properties within the domain. FEM is frequently used in mechanical and thermal engineering disciplines. It has also been applied to electromagnetic field analysis with many commercial RF simulations tools available in the market that provide significant advantages over analytical methods.

4.2.4.2 Method of Moments

The Method of Moments transforms integrated and differential equations into matrix systems of linear equations that can be solved using computers. MoM divides the integration domain into small elements with the assumption that the unknown does not vary significantly over each elementary cell [8]. The elements are small compared to the wavelength of interest. Each of these segments will carry some current, and the current on each segment will affect the current on every other segment. The unknown function is approximated by a finite series of known expansion functions with unknown expansion coefficients. This results in a number of simultaneous algebraic equations for the unknown coefficients. These equations are then solved using matrix calculus [9]. Far field and near field can be calculated and presented as an entire complex structure. Maximum field strength patterns can be observed as well as the interaction of radiation patterns, RF current distribution, and the radiating-source impedance [10]. The Method of Moments has the advantage of being relatively simple to implement. Because it is a frequency domain technique, MoM is inefficient to use over a larger frequency range.

4.2.4.3 Finite-Difference Time-Domain

FDTD time-stepping relation assumes that, at any point in space, the updated value of the electrical field in time is dependent on the stored value of the E-field and the numerical curl of the local distribution of the magnetic field in space [11]. As a time-domain solution, FDTD uses a grid-based differential numerical analysis technique, where the grids are small compared to the smallest wavelength of interest. The entire surrounding space has to be gridded in square or rectangular units for 2D modeling, or cubic structures for 3D modeling.

FDTD provides a wide frequency range solution with a single simulation run, with full consideration to nonlinear material properties. It solves the electrical field and magnetic fields at any point within the domain by modifying Maxwell's equations to fit a computational algorithm. The FDTD method is one which is very well suited to the modeling of the propagation of electromagnetic fields through three-dimensional volumes containing materials of differing permeability (μ), permittivity (ε), and conductivity (σ) [12]. It allows the user to specify the material at all points within the computational domain. The FDTD technique also differs from the Methods of Moments in that the entire space of interest needs to be gridded, which requires very large computational and very long solution times.

4.2.4.4 Finite-Differences Frequency-Domain

FDFD is a numerical solution method based on finite-difference approximations used for electromagnetic radiation simulation and is excellent for field visualization. It transforms Maxwell's equations for sources and fields at a constant frequency into a set of linear algebraic equations in matrix formwith no time steps. The FDFD algorithm is easy to implement but it requires very large computational power to solve a sparse matrix. It can be used to minimize power incident on absorbing boundary conditions and to automatically separate the source from scattered waves for easier post-processing. FDFD is able to resolve sharp resonances and obtain solutions at a single frequency [13]. However, it is ill suited to solve multiscale structures and complex geometries due to its inability to scale well to three dimensions.

4.3 Shielding Techniques and Methods

An electromagnetic shield structure for RF and microwave SiPs and other packages is based on a Faraday's cage formed by a conductive material. The conductive metallic cage has to be well grounded with very low resistive contact area to stop the penetration of the electromagnetic field into the SiP's shielding wall and block energy radiation to the outside field. The shielding performance changes with frequency or faster variation of the electric field due to skin depth of the conductive shielding material. Higher frequencies increase the wave leakage from inside the SiP through cuts, slits, and meshes in the conductive layer of the shielding material and at the substrate layer of the SiP through vias, anti-vias, ground openings, and interconnects, such as land grid array (LGA) pads or ball grid array (BGA) balls.

Shielding performance is directly proportional to the shielding material's electrical conductivity. Other factors impacting shielding performance are shielding wall thicknesses, frequency, skin depth, the size of the shielded area and volume, how the shielding metal is connected to the SiP, and mother board ground layers structure.

4.3.1 Metal Caps

Conventionally, metallic caps are employed to prevent EM noise at the package level. SiPs and other package designs with metallic caps, however, suffer from high cost and large size—even though the caps offer excellent shielding effectiveness and good mechanical protection. For this design, a thin flat piece of metal is cut and formed into a variety of shapes by a metal shop. Copper is generally used for RF and microwave shielding due to its ability to absorb electromagnetic waves.

The metal cap can be either solid metal or metal with holes in mesh or non-mesh forms. Solid metal shielding can either have solid metal on the top and four sides of the cap, or solid on the top of the cap and notches on the four sides for air circulation and a moisture escape path (see Fig. 4.5a). Metal shielding with mesh holes are used in SiPs that contain active and passive components with molding requirements. The holes and cuts in the shield must be significantly smaller than the wavelengths of the frequencies of interest and large enough to flow mold compound materials through the holes and cuts without causing any mold defects or voids as shown in Fig. 4.5b, c.

Depending on the assembly design rules, A + B + C = 0.300 mm–0.750 mm identifies the additional space required for the sides of the SiP's substrate, and D and E (0.175 mm – 0.350 mm) depict the increase in the SiP's thickness. For example, the size of 10 mm x 10 mm (100 mm^2) footprint SiP increases to 11 mm x 11 mm,

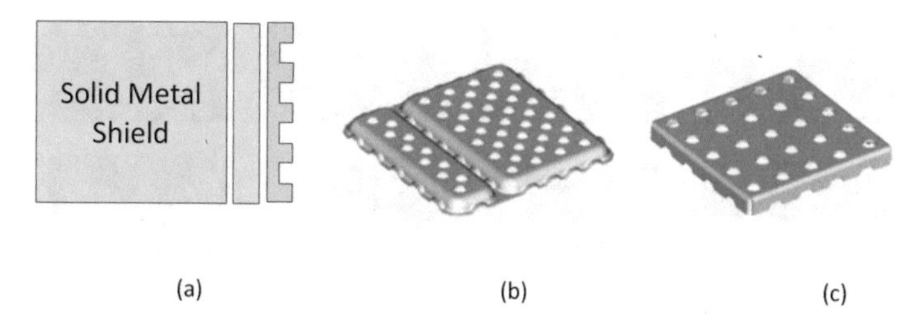

Fig. 4.5 (**a**) Solid metal, (**b**) formed metal shield with holes, and (**c**) folded metal shield with holes

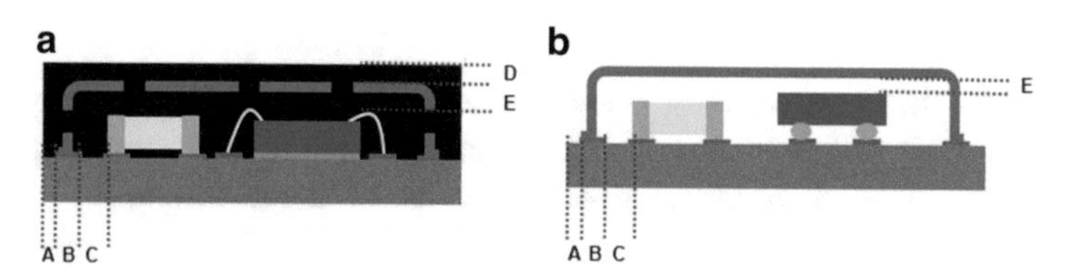

Fig. 4.6 (**a**) SiP embedded shielding with wire bond die and molding. Metal shielding requires additional space in x, y, and z directions and (**b**) SiP metal shielding without mold compound

equivalent to additional 21 mm^2 increase in a package footprint with additional cost penalties.

4.3.2 *Plating*

The metal plating process involves creating an outer metal coating layer using copper, nickel, or other conductive metals. It is usually performed by immersing the metal in an acid solution with an anode electric current and cathode. The material to be plated is the cathode (negative electrode) of an electrolysis cell that conducts a direct electric current [14]. The plating process for the RF SiP requires coating the mold surface. To do this, the surface is chemically roughened to support the adhesion of an electroless Cu as a seed layer to carry current for remaining actual shield layer. The main shielding layer is electrolytic copper, but a thin layer of nickel may be required for environmental protection of the copper. The plating process (Fig. 4.7) starts by masking the active sides of the SiP package to prevent plating on functional areas, and the mask is removed prior to the final stage of device singulation.

Morris describes the process as “masking can be done with tape or a fixture, and is an important operation in the electroless plating process because all exposed areas are plated during this step [15].”

The plating technique has potential for good thickness control and full coverage of the SiP. It also has a high through puts units per hour (UPH) rate at the strip level,

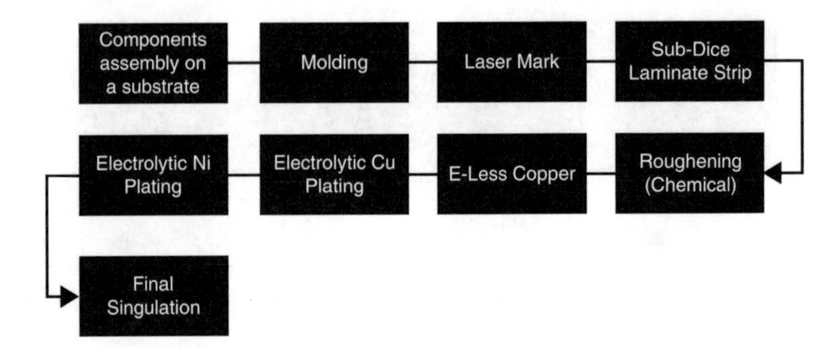

Fig. 4.7 The steps of a typical plating process

good yield and low material cost. However, the plating process requires large, complex facilities. The other disadvantages for the plating are additional surface pretreatment, complicated masking, wet processes, and water treatments.

4.3.3 Spray Coating

Spray coatings have been developed to add shielding characteristics to plastic enclosures and package molding compounds to protect electronic circuits within a package from electromagnetic interference.

In this process, nontoxic gas air or nitrogen gas is atomized into a spray for a polymer solution creating fine mist to sweep across the surface of a package's mold compound. The electrically charged atomic mist lands on the top and four sides of the package covering the exposed substrate ground layer from all sides of the package. This forms a Faraday cage with metal coverage of the top and four sides of the package and its connection to the package's substrate ground. Process variables include solution viscosity and solid contain, solution flow rate, nozzle size and distance from target device, nozzle pressure, atomization pressure, and sweep speed [16].

Electrostatic spray systems are available in manual or automatic gun models. An automatic electrostatic spray system is used in smaller scale applications.

4.3.4 Sputtering

Sputtering is a technique used to deposit a thin film of copper, silver, NiFe, and other conductive materials onto the surface of the package's mold compound. The sputtering process is performed by ionizing inert gas particles in an electric field and then directing the resulting gas plasma toward the target package [17]. The sputtering particles produce thin films of metal on the top and four sides of the package's mold compound and connect to the package substrate ground similar to the spray

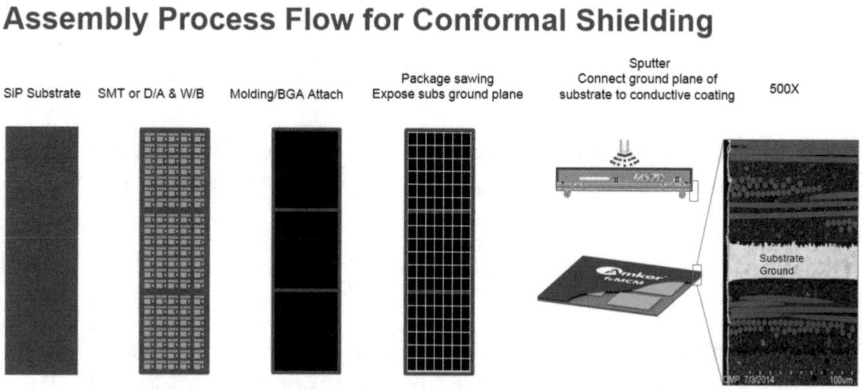

Fig. 4.8 Assembly process flow for conformal shielding

techniques described in Sect. 4.3.3. As shown in Fig. 4.8, the conformal shielding process flow for an SiP design starts with an SiP substrate strip, adds SMT, die attach (DA) wire bond, molding, BGA attach and package sawing for exposing the substrate ground plane, then sputtering conductive metal to connect the substrate ground to the conductive coating to form a Faraday cage. Sputtering temperature, pressure, and power are used by equipment manufacturers to control physical properties of the sputtering uniformity, adhesion, and alloy composition.

4.4 Shielding Performance of Thin-layer Conformal Shielding

Conformal shielding techniques have been used for RF package shielding to reduce packaging thickness, increase production yield, and enhance overall shielding performance of a package. The shielding effectiveness (SE) value depends on the shielding material, its thickness, size of the shielded volume, grounding structures, and the frequency of interest.

Due to the wavelength's attenuation to $1/e$ at one skin depth, the multi-reflection losses can be ignored for the cases, where the shielding thickness is much larger than the skin depth. To summarize, shielding effectiveness:

$$SE_{dB} = R_{dB} + A_{dB} + M_{dB} \tag{4.2}$$

where R_{dB} is the reflection loss, A_{dB} is the absorption loss, and M_{dB} is the multi-reflection loss. C. R. Paul estimates the reflection loss of a conductor with a uniform plane wave as [18]:

$$R_{dB} = 20 log_{10} \left| \frac{(\eta_0 + \eta)^2}{4\eta_0 \eta} \right| \tag{4.3}$$

where η_0 and η are the wave impedance in free space and the conductor, respectively. For a good conductor, $\eta << \eta_0$, so R_{dB} can be further expressed as [18]:

$$R_{dB} \cong 20 log_{10} \left| \frac{\eta_0}{4\eta} \right| \tag{4.4}$$

The absorption loss is the attenuation of the electromagnetic wave when it propagates inside the shielding material. It can be mathematically expressed as [18]:

$$A_{dB} = 20 log_{10} e^{\frac{t}{\delta}} \approx 8.69 \frac{t}{\delta} \tag{4.5}$$

where t is the thickness of the shielding material, and δ is the skin depth. The multi-reflection losses can be expressed as [18]:

$$M_{dB} \approx 20 log_{10} \left| 1 - e^{\frac{-2t}{\delta}} \right| \tag{4.6}$$

If the thickness of the shielding material is greater than the skin depth at the operating frequency, the transmitted EM wave attenuates very quickly during its propagation inside the shielding material, and the multi-reflection loss can be ignored [19].

4.4.1 *Shielding Performance Measurement Methods*

The electromagnetic varying field is typically divided into near field and far field. The common definition of the near field is the source is less than one wavelength away from the observation point antenna. The wavelength is given by Eq. (4.1) in Sect. 4.2.2.

The near field is divided into two areas: the reactive and the radiative. In the reactive area, where the electrical and magnetic fields are the strongest, both fields can be measured separately with two different antennas or probes. The magnetic (H) field is usually measured by a loop antenna [20].

The second region of the near field is the radiative region. In the radiative area, the fields begin to radiate. This area represents the beginning of the far field. In the near field, the strength of the fields varies inversely with the cube of the distance from the antenna ($1/r^3$) [20]. Near-field scanning may provide information about the surface current, tangential fields, and the reactive near-field distribution on the device under test (DUT) [21].

The far-field region extends more than two wavelengths away from the radiation source. The electromagnetic field at far field has the characteristics of a plane wave, so only the electrical field needs to be measured in an isolated anechoic or semi-anechoic chamber.

4.4.2 *Test Vehicles*

Most RF SiPs are based on LGAs instead of BGAs to reduce capacitive and inductive parasitics and to minimize thermal resistance to reduce the thermal path especially for power amplifier (PA) SiPs. Amkor Technology with Missouri University of Science and Technology Lab jointly conducted a series of tests of conformal shielding performance through a wide range of frequency bands from 100 MHz to 12 GHz.

The test vehicle (TV) in the experiment is an LGA module mounted on an interposer which is used to connect to the test equipment for lab testing. The TV module is a 4-layer substrate within an LGA package, which includes a broadband antenna on the top layer (L1), a reference layer (L2), ground layer (L3), and LGA pads layer (L4). The interposer is a 2-layer board that consists of ground and LGA pads (L1), and an antenna feed and co-planar ground (L2). Two types of TV modules were built: an unshielded module as a reference and a module with a sputtered, conformal shield. The sputtered shields are 3 μm, 5 μm, and 7 μm thick for studying EMI performance due to shielding wall thickness. The size of the TV module is 10 mm x 10 mm x 1.2 mm (Fig. 4.9a) and the dimensions of the interposer are 40 mm x 25 mm x 0.47 mm Fig. 4.9b [22].

The proposed conformal shield employs sputtered conductive material onto the package mold. The packages to be shielded are assembled and molded using the assembly flow described in Sect. 4.3.4.

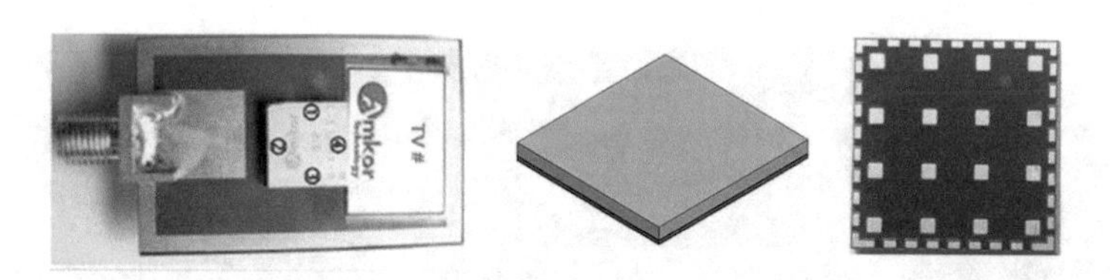

Fig. 4.9 (**a**) Interposer board and TV sample and (**b**) TV sample. Courtesy of Amkor Technology

4.4.3 Shielding Effectiveness

The SE of a shield is often defined as the ratio of the electric fields at any arbitrary point between an unshielded package ($E_{\text{unshielded}}$) and a shielded package (E_{shielded}), and usually expressed in decibels as:

$$SE(dB) = 20log_{10}E_{\text{unshielded}} - 20log_{10}E_{\text{shielded}} \tag{4.7}$$

4.4.4 Far-field Shielding Performance Measurements

A vector network analyzer (VNA) was used in far-field testing to measure the S-parameter of the test setup chain, specifically S21. In this case, for far-field measurements, SE can be expressed as:

$$SE(dB) = S_{21}^{u}(dB) - S_{21}^{s}(dB) \tag{4.8}$$

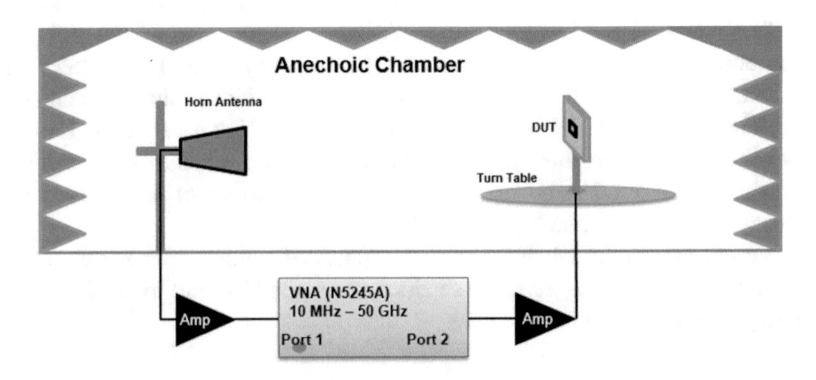

Fig. 4.10 Block diagram of the far-field measurement test setup

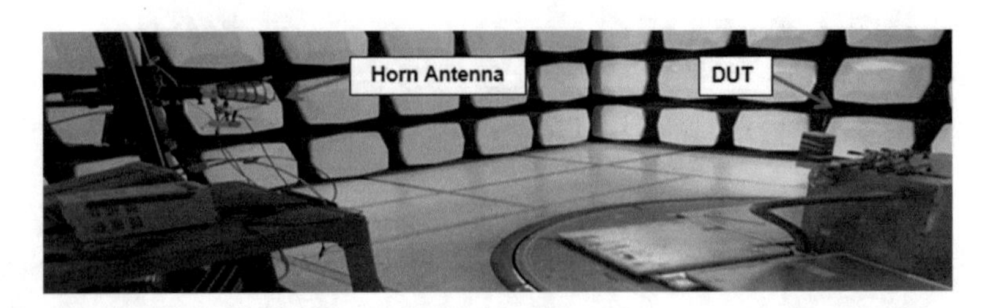

Fig. 4.11 Semi-anechoic chambers lab for far-field electromagnetic measurements. Courtesy of Missouri University of Science and Technology

where

S_{21}^{u} and S_{21}^{s} are S_{21} from unshielded and shielded devices under test (DUTs), respectively, as illustrated in Fig. 4.10.

The far-field test setup and lab photos are shown in Figs. 4.10 and 4.11, respectively. Far field is measured from 0.8 GHz to 12 GHz.

It is difficult to predict the exact location of the maximum radiation area from the RF SiP. To detect and record the maximum radiated emission in 3D format, the receiving antenna must be set to both horizontal and vertical polarizations and the turn-table has to be rotated by 360°. In addition, the DUT has to move up and down vertically by 12–36 in. Collected data must be processed to flag the maximum radiated emission data needed to calculate the shielding performance. Extreme caution must be taken to minimize the noise floor in the anechoic chamber by minimizing noise from signal sources, amplifiers, cables, and other equipment. Ferrite chokes, wave-absorbing materials, and good grounding techniques are common practices to improve the measurement integrity.

In far-field tests, radiation from four different TV samples, i.e., unshielded module, sputtered wall thickness of 3 μm, 5 μm, and 7 μm modules were recorded from 0.8 to 12 GHz. The far-field SE data, calculated based on measured data using Eq. (4.8), are shown in Fig. 4.12. Clearly, the SEs of all shielded modules are around 40 dB within the entire band from 0.8 to 12 GHz, excellent performance overall [22].

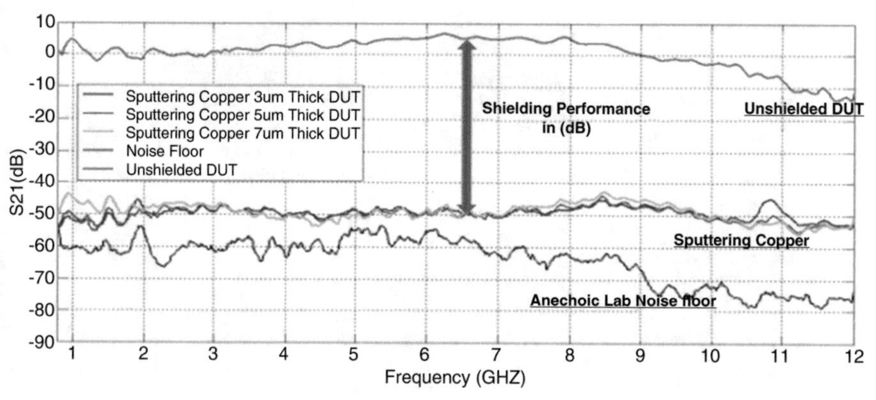

Fig. 4.12 Far-field measurement data for unshielded and shielded SiPs using 3 μm, 5 μm, and 7 μm copper sputtering wall thicknesses. Courtesy of Amkor Technology

4.4.5 Near-field Shielding Performance Measurements

For near-field measurements, both the H-field and the E-field are measured. The E-Field and H-Field shielding effectiveness are expressed as:

$$\mathrm{SE}(\mathrm{dB}) = \mathrm{E}^{\tan u}(\mathrm{dBm}) - \mathrm{E}^{\tan s}(\mathrm{dBm}) \quad (4.9)$$

$$\mathrm{SE}(\mathrm{dB}) = \mathrm{H}^{\tan u}(\mathrm{dBm}) - \mathrm{H}^{\tan s}(\mathrm{dBm}) \quad (4.10)$$

where $E_t = \sqrt{\left(E_x^2 + E_y^2\right)}$ is the tangential E-field, and E_x and E_y are measured power with an E-Probe in dBm (decibel milliwatts); $H_{\tan} = \sqrt{\left(H_x^2 + H_y^2\right)}$ is the tangential H-field and H_x and H_y are measured power with an H-Probe in dBm; $E^{\tan u}$ and $E^{\tan s}$ are the tangential E-field from unshielded and shielded DUT; and $H^{\tan u}$ and $H^{\tan s}$ are the tangential H-field from the unshielded and the shielded DUT.

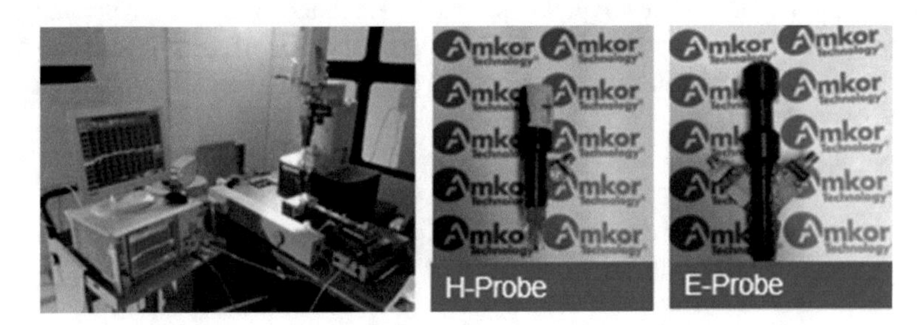

Fig. 4.13 Near-field measurement setup. Courtesy of Amkor Technology

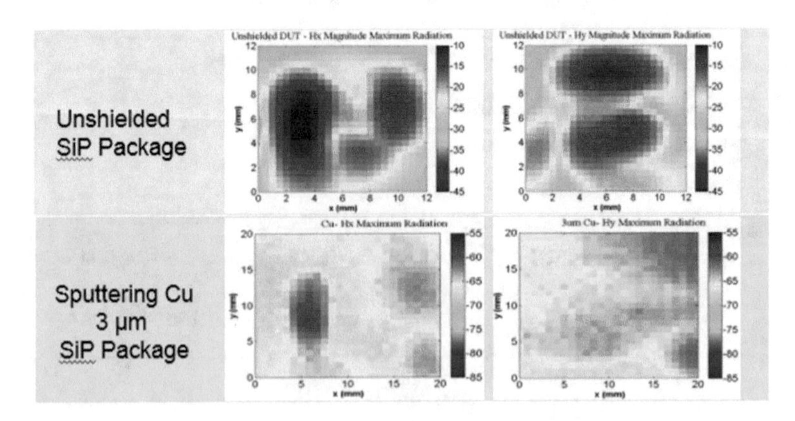

Fig. 4.14 Maximum near-field radiation (0.1–6 GHz): DUT-Hx, Hy. Courtesy of Amkor Technology

The near-field test setups and lab photos and probes are shown in Fig. 4.13. Near field is scanned from 0.1 GHz to 6 GHz. For near field, both the E-field and H-field are measured using an E probe and an H probe, respectively.

The near-field measurements were performed using a scanning system on a probe station as shown in Fig. 4.13. Both E-field and H-field probes were employed to pick up radiation in the near-field region. The scanned area is 12 mm x 12 mm, with a step size of 0.5 mm along both the x and y directions. There are 24 rows and 24 columns, resulting in a total of 576 measurement points at each frequency. The frequency range for the near-field test is 100 MHz–6 GHz.

Some of the near-field measured data are shown in Fig. 4.14, which displays the radiated E-field patterns along both the x direction (|Ex|) and the y direction (|Ey|) for unshielded and 3 μm sputtered samples. H-field patterns indicate similar performance differences, but are not shown here. Note that for picture clarity, unshielded and shielded field patterns use different scales.

The E- and H-field rainbow patterns in Fig. 4.14 clearly reveal that the sputtered shield provides significant shielding effectiveness for near-field E and H radiation [19].

4.5 Summary

This chapter introduced the need for shielding electronic packaging devices. It explained electromagnetic wave radiation and the sources of the radiation with details regarding why an antenna radiates electrical and magnetic fields. The theoretical view on how charged particles' acceleration produces electromagnetic waves, and how Maxwell's equations define electromagnetic fields with relation to current and voltage potential were explained. To identify radiation sources and radiation patterns in a system, system designers must understand how a particular simulation tool solves Maxwell's equations. Subsequently, different methods to solve Maxwell's electromagnetic equations to predict the wave propagation were discussed.

Conformal shielding in package and System-in-Package design considerations, analyses, and methods to approach optimum solutions were presented. Different SiP and package-level shielding techniques and methods were covered in detail with emphasis on conformal shielding. Finally, far-field and near-field measurement techniques to measure SiP and package-level shielding performances were proposed.

Acknowledgement The authors wish to thank the EMC Laboratory, Missouri University of Science and Technology, and Rong Zhou, Jingun Mao, JinSeong Kim and Ben Zarkoob at Amkor Technology, for building the test samples, technical support, and measurements.

References

1. Greenberry, Eve. *An activist's journey to raise awareness about electromagnetic pollution part 1, part 2*, http://electromagnetichealth.org/wp-content/uploads/2011/01/Explore-Part-I-Final-with-OK.pdf.
2. http://ec.europa.eu/growth/sectors/electrical-engineering/emc-directive/.
3. Kenneth Keenan, R. 1983. *Digital design for interference specifications*.
4. Dash, Glen. *EMI: why digital devices radiate*, http://glendash.com/downloads/electromagnetics/as-published/Why_Devices_Radiate_AP.pdf, pp. 4–7.
5. Hofhiens, Jared. *Antennas an introductory guide*, http://www.idc-online.com/technical_references/pdfs/electronic_engineering/Antennas_an%20Introductory%20Guide_20061128.pdf, p. 7.
6. Dash, Glen. *A Dash of Maxwell's: a Maxwell's Equations Primer, Part Two*, pp. 2–10.
7. Fujimoto, Kyohei, and Morishita Hisashi. *Modern Small Antennas*, Cambridge University Press, p. 389.
8. Massachusetts Institute of Technology, *The Method of Moments in Electromagnetics*, 6.635 lecture notes.
9. A tutorial on the method of moments. 2012. *IEEE Antennas and Propagation Magazine* 54(3): 260–275. https://www.researchgate.net/publication/260515353_A_Tutorial_on_the_Method_of_Moments.
10. *Software analysis simulation methods*, RADIOING.com—eEngineer, http://www.radioing.com/eengineer/methods.html.
11. Yee, Kane. 1966. Numerical solution of initial boundary value problems involving Maxwell's equations in isotropic media. *IEEE Transactions on Antennas and Propagation* 14(3): 302–307.
12. Dash, Glen. *Computational magic and the EMC engineer*, http://glendash.com/Dash_of_EMC/Computational_Magic/Computational_Magic.pdf.
13. Rumpf, R.C. 2012. Simple implementation of arbitrarily shaped total-field/scattered-field regions in finite-difference frequency-domain. *Progress in Electromagnetics Research B* 36: 221–248.
14. *Guide to metal plating*. http://www.weldguru.com/metal-plating.html.
15. Morris, Scott, and Schonthal Eric. *Shielding RF components at the package level*, https://hubslide.com/rrpipkin/wp-shielding-rf-components-at-the-package-level-2-s5758263f390ab8e8461b2e0d.html.
16. Tummala, Rao R., et al. 1997. *Microelectronics packaging handbook*. 2nd ed, 674. Springer.
17. Christiansen, Donald. 1996. *Electronics Engineers' Handbook*, 4th edition. McGraw-Hill. Chapter 11. p. 25.
18. Paul, C.R. 1992. *Introduction to electromagnetic compatibility*. A Wiley-Interscience Publication.
19. Karim, N., J. Mao, and J. Fan. 2010. Improving electromagnetic compatibility performance of packages and SiP modules using a conformal shielding solution, *2010 Asia-Pacific Symposium on Electromagnetic Compatibility*, Beijing.
20. Frenzel, Lou. 2012. What's the difference between EM near field and far field?, *Electronic design*. http://electronicdesign.com/energy/what-s-difference-between-em-near-field-and-far-field#2.
21. He, Hui. *The development of near field probing systems for EMC near field visualization and EMI source localization, summer 2015*, Missouri S&T, Rolla, MO.
22. Karim, Nozad, Rong, Zhou, and Jun, Fan. An innovative package EMC solution using a highly cost-effective sputtered conformal shield. In *IMAPS2016, 12th International Conference and Exhibition on Device Packaging*, Fountain Hills, AZ, March 2016.

Chapter 5
Design of C-Band Interdigital Filter and Compact C-Band Hairpin Bandpass Film Filter on Thin Film Substrate

Min Tan, Yang Xuan, Yong Ma, Li Li, and Yonghe Zhuang

5.1 Introduction

With the development of modern electronic technology, microwave communication equipment tends to miniaturization, low cost, and high performance. While microwave integrated circuit modules become more and more abundant, the discrete circuit applications will be fewer and fewer [1]. Now microwave integration technology has become mature, and microstrip and microstrip circuit are widely used in microwave integrated circuits due to small size, light weight, low cost, and so on.

This paper studies the C-band microstrip interdigital and hairpin filter, which is used in the transceiver channel. The C-band filter, which is soldered to the PCB board, could reduce the impact of spurious signals in the transceiver. The frequency source is an important part of the transceiver, which provides the local oscillator signal to the transceiver channels. The spurious signals of the transceiver system require less than 60 dBc. The experimental filter of this paper is fabricated on an Al_2O_3 ceramic substrate with a relative dielectric constant of 9.8, thickness of 0.381 mm and the PCB board is fabricated on a commercial substrate (Rogers R04350) with a relative dielectric constant of 3.66, and loss tangent of 0.004 more than Al_2O_3 ceramic. The first filter is designed in the PCB board with thickness of 0.381 mm.

The proposed filter of this paper has a better performance than the first filter. First, the precision of processing of the first filter designed in the PCB board is worse than Al_2O_3 ceramic substrate, and the loss tangent of Rogers R04350 is bigger than Al_2O_3 ceramic. Second, because the design of frequency source is complex and the PCB board is multilayer, the bottom of the original filter unpaved lot of ground

M. Tan (✉) • Y. Xuan • Y. Ma • L. Li • Y. Zhuang
East China Research Institute of Microelectronics, Hefei 230022, China
e-mail: shuiyue4149@126.com

K. Kuang, R. Sturdivant (eds.), *RF and Microwave Microelectronics Packaging II*,
DOI 10.1007/978-3-319-51697-4_5

and only around the filter is grounded, so the ground is not very good, and performance is less than the indicator of the design.

The size of proposed filters is relatively smaller than the first filter because of the dielectric constant, and consistency in the high and low temperature characteristics is also very good because of the advanced technology of the film. In order to realize miniaturization, film substrate is used to design the filter and simulation is conducted by means of the electromagnetic simulation software. At the end of the paper, the vector network analyzer is used to measure the filter, and the test results show that the performance of the filter reaches usage requirements.

5.2 Microstrip Filter

The filter is based on a resonant circuit, which is a two-port network, the frequency signal within the passband can transmit, and within the stopband the signal is attenuated because of mismatch. The filter can achieve spectrum shift function.

The microstrip filter has the features of a simple design, easy to produce, small size, low cost, and ease of integration, and is used in a very wide range in microwave integrated circuits [2, 3].

The common structure of microstrip filter has hairpin, parallel linear coupling, and interdigital structure etc. [4, 5]. The interdigital filter is defined by the parallel coupled line resonator which consists of a cross structure, as shown in Fig. 5.1 [4]. It has the following advantages: compact structure, high reliability, lower tolerances between each resonator, and easy to manufacture.

Hairpin-line bandpass filters are compact structures. They may conceptually be obtained by folding the resonators of parallel-coupled, half-wavelength resonator filters, which were discussed in the previous section, into a "U" shape. This type of "U" shape resonator is the so-called hairpin resonator.

The C-band hairpin bandpass is designed to have a fractional bandwidth of 15% at a midband frequency $f_0 = 8$ GHz. A five-pole ($n = 5$) Chebyshev lowpass prototype with a passband ripple of 0.1 dB is chosen for the first filter. The lowpass prototype parameters, given for a normalized lowpass cut of frequency $c = 1$, are $g_0 = g_6 = 1.0$, $g_1 = g_5 = 1.1468$, $g_2 = g_4 = 1.3712$, and $g_3 = 1.9750$. The bandpass design parameters can be calculated by

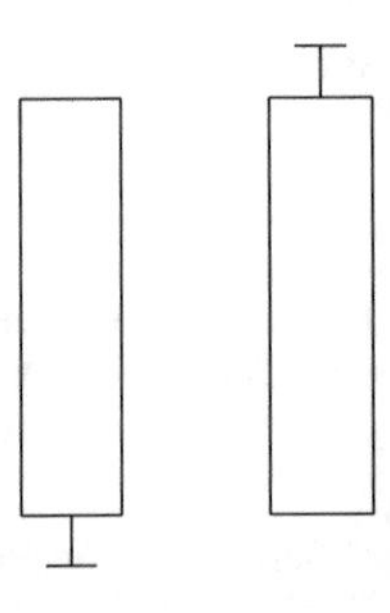

Fig. 5.1. The general form of interdigital filter

$$Q_{e1} = \frac{g_0 g_1}{FBW}, Q_{en} = \frac{g_n g_{n+1}}{FBW}$$

$$M_{i,i+1} = \frac{FBW}{\sqrt{g_i g_{i+1}}} 11 \le i \le n-1$$

5.3 Design of the Interdigital Filter

For integration and miniaturization, the interdigital structure is used to design the filter presented here.

5.3.1 *Structure of the Filter*

The filter in the paper is used in the transceiver channel. Film substrate of the back metallization is used to facilitate the assembly The film substrate has high precision, high dielectric constant, and low loss characteristics, and can be used to realize the miniaturization of the interdigital filter. In this paper, we select a kind of film substrate whose thickness is 0.381 mm. The dielectric constant of film substrate is 9.8, and the thickness of the gold layer on the surface is 3 um.

Figure 5.2 shows the structure of the bandpass interdigital filter, which is a vertically symmetrical structure. The filter consists of four L-shaped microstrip lines, which are composed of a short-circuited microstrip line at one end and the other end is open. The short-circuited end is connected to ground through the hole formation.

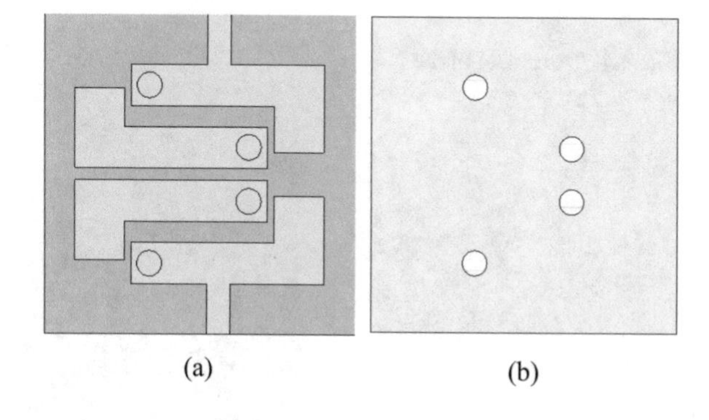

Fig. 5.2 The structure of the filter: (**a**) top view, (**b**) bottom view

5.3.2 The Simulation Analysis

We use the electromagnetic simulation software to simulate the filter in Fig. 5.2, and Fig. 5.3 shows the main variable parameters of the structure.

The value of optimized parameters are as follows:

$W_1 = 0.8$ mm,
$W_2 = 1.3$ mm,
$L_1 = 2.1$ mm,
$L_2 = 3.2$ mm.

The model is shown in Fig. 5.4, and uses the optimized parameters. Figure 5.5 shows the final *S*-parameter simulation results.

As can be seen from Fig. 5.5, the insertion loss is less than 1 dB in the center of the band, and out of band rejection (12–18 GHz) is more than 50 dB, so the results meet the design requirements.

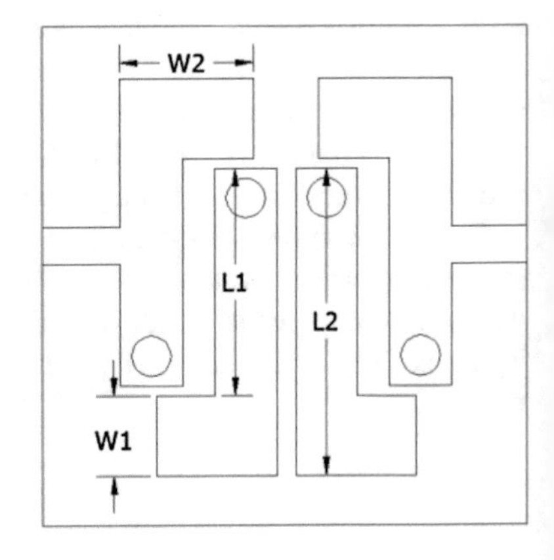

Fig. 5.3 The structure parameters

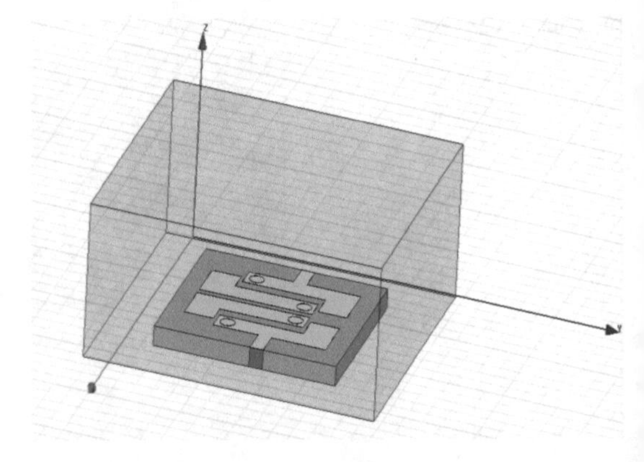

Fig. 5.4 The model of the filter

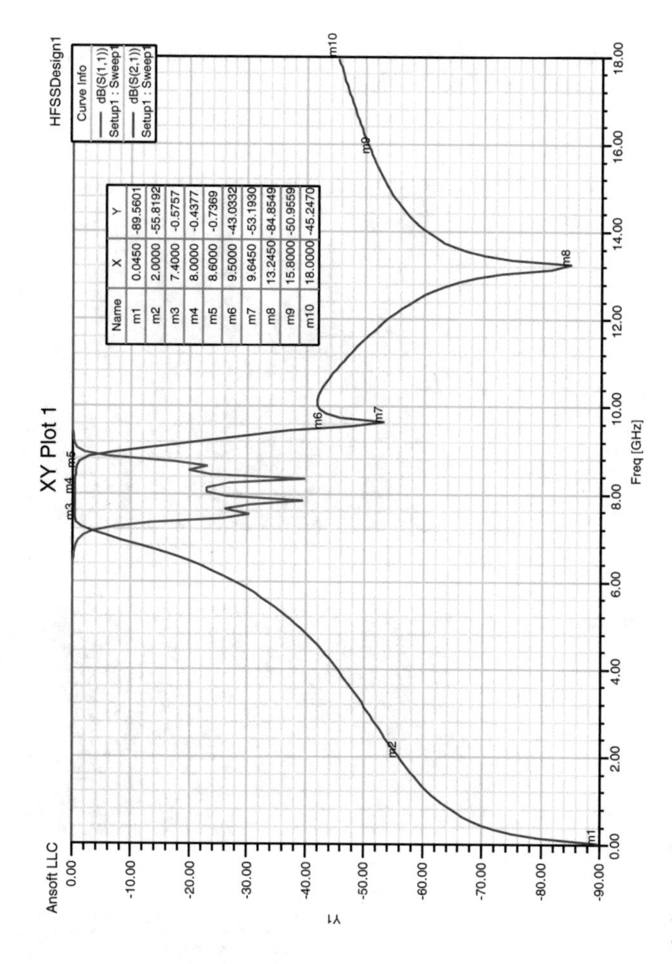

Fig. 5.5 S-parameter simulation results

5.3.3 The Test Results

Figure 5.6 shows the bandpass interdigital filter made of the film substrate, and the actual size is 5 mm × 5 mm × 0.381 mm.

The test fixture is used to test the performance of the filter as shown in Fig. 5.7.

The indicators of the filter are measured by means of vector network analyzer, as shown in Fig. 5.8.

The test results show that the insertion loss is about 1.5 dB and out of band rejection is about 45 dB, which conforms to the requirements of the indicator, and the filter meets the requirements of the channel.

Fig. 5.6 Picture of the bandpass interdigital filter

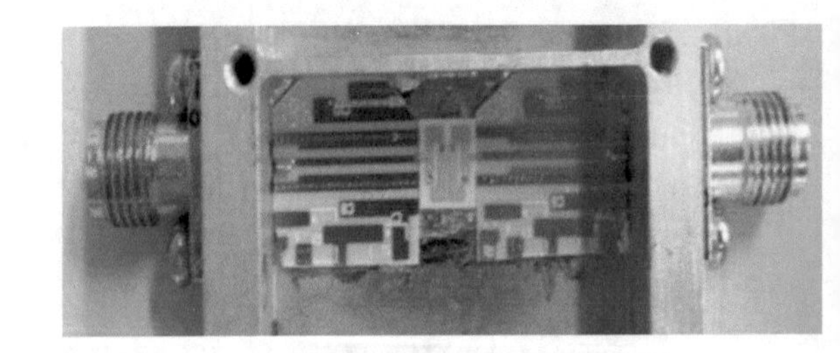

Fig. 5.7 Picture of test fixture

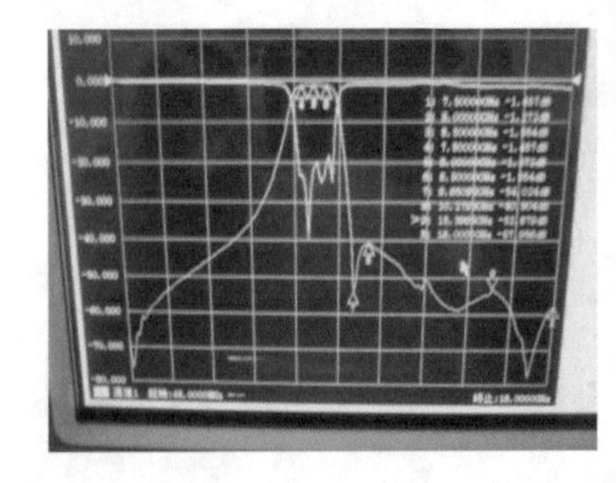

Fig. 5.8 The test results

5.4 Design of the Hairpin Filter

5.4.1 *Structure of the Filter*

We use a commercial substrate (Rogers R04350) with a relative dielectric constant of 3.66 and a thickness of 0.254 mm for the design of the first filter. We then carry out full-wave EM simulations to extract the external Q and coupling coefficient M against the physical dimensions. At last, we can obtain the characteristics of the filter:

$W_0 = 0.2$ mm,
$W_1 = 0.525$ mm,
$W_2 = 1.3$ mm,
$L_1 = 5.6$ mm,
$L_2 = 5.1$ mm,
$S_1 = 0.12$ mm,
$S_2 = 0.18$ mm,
$t = 1.6$ mm

Figure 5.9 shows the layout of the origin 5th-order C-band hairpin bandpass filter, which is a vertically symmetrical structure.

5.4.2 *The Simulation Analysis*

Figure 5.10 shows the full-wave simulated performance of the origin filter.

Because the performance of the origin 5th-order C-band hairpin bandpass filter is less than the indicator of the design, we designed the proposed 7th-order C-band hairpin bandpass filters. Figure 5.11 shows the layout of the 7th-order C-band hairpin bandpass filter, which is a vertically symmetrical structure. Figure 5.4 shows the proposed filter soldered on the PCB board. The green part of the figure represents

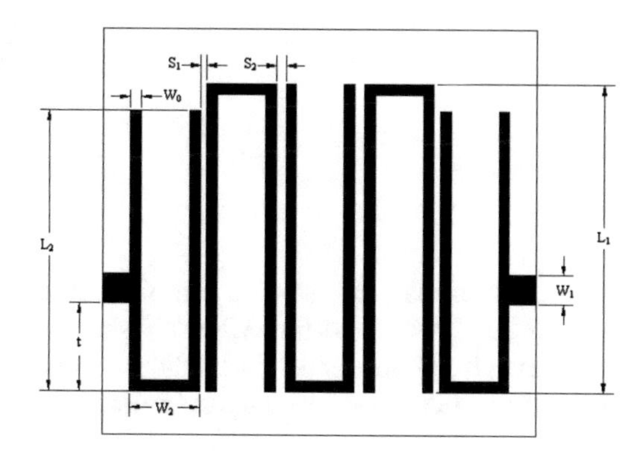

Fig. 5.9 Layout of the origin filter

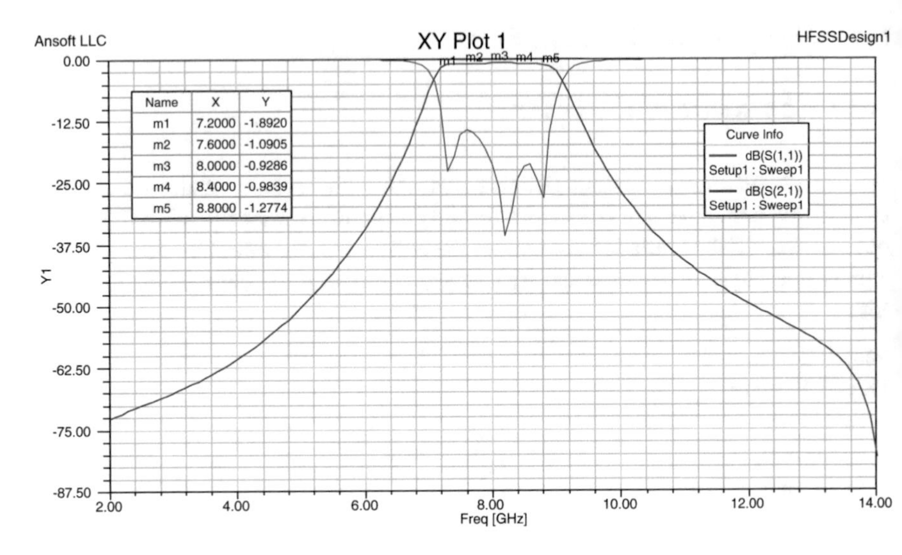

Fig. 5.10 Full-wave simulated performance of the origin filter

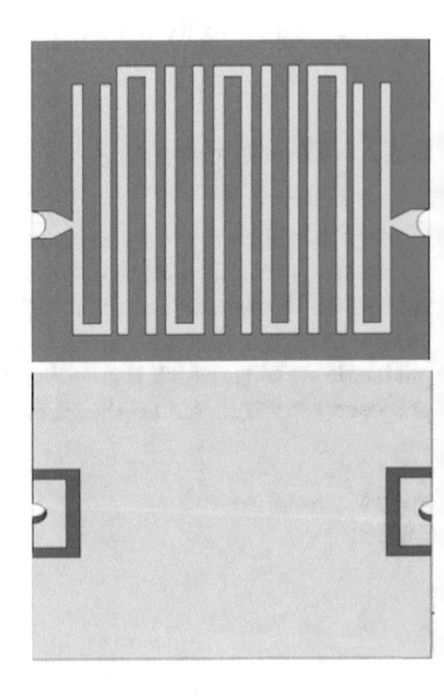

Fig. 5.11 Layout of the proposed filter

the PCB substrate, and the blue part of the figure represents the 7th-order C-band hairpin. The filter is soldered to the PCB substrate. From the picture we can find that the pads in the bottom of the filter, which are connected to the input and output ports using the vertical vias, are compact in size.

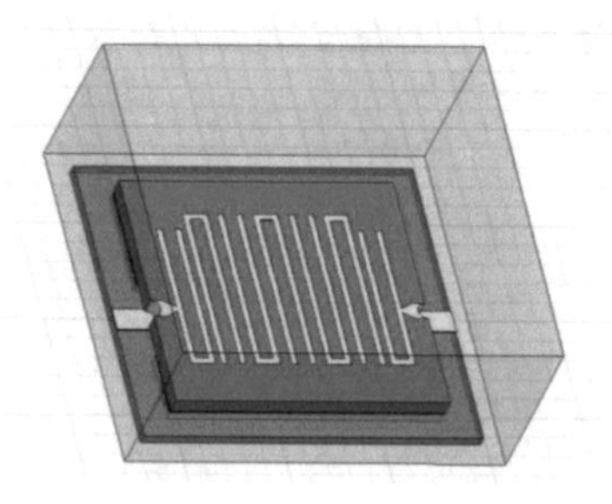

Fig. 5.12 Layout of the proposed filter soldered on the PCB board

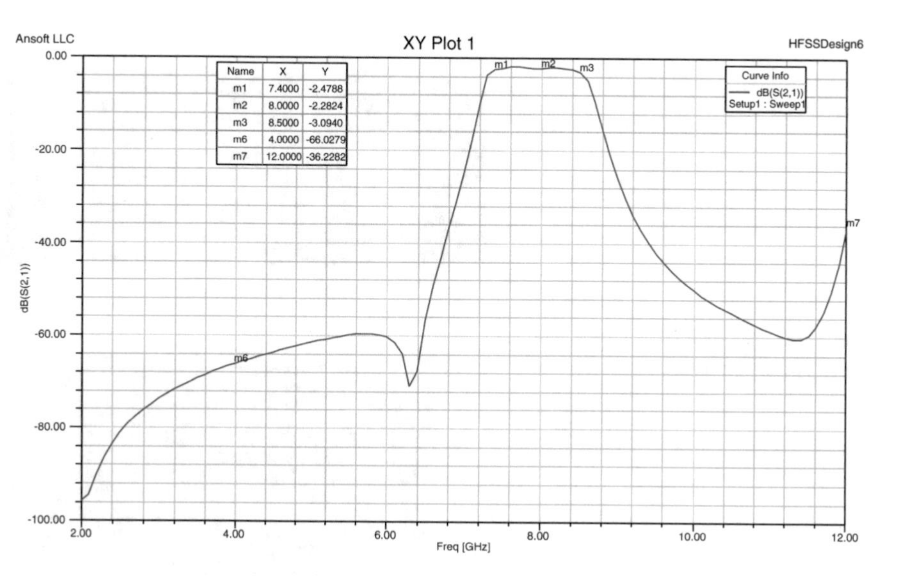

Fig. 5.13 Full-wave simulated performance of the proposed filter

Figure 5.13 shows the full-wave simulated performance of the origin filter. As can be seen from Fig. 5.13, insertion loss in the center of the band is less than 3 dB, and band rejection is at about 30 dB. It basically meets the requirements of the design of filter.

5.4.3 *The Test Results*

Figure 5.14 shows the layout of the 7th-order C-band hairpin bandpass filter and the actual size is 7 mm by 7 mm by 0.381 mm. As can be seen from Fig. 5.5, the filter is made of the film substrate. The test fixture is used in order to test the performance of the filter, as shown in Fig. 5.15.

Fig. 5.14 Picture of the proposed filter

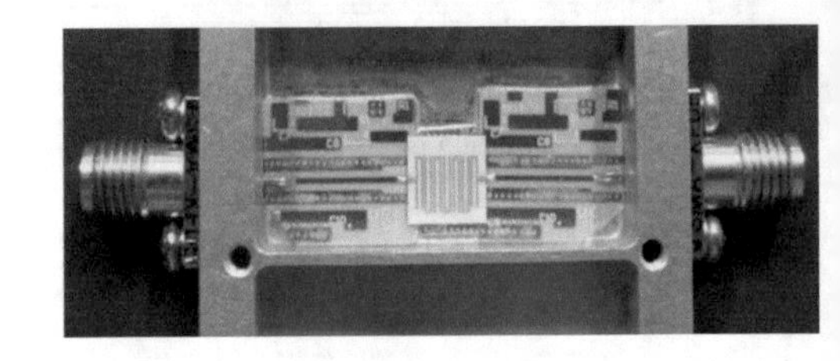

Fig. 5.15 Picture of test fixture

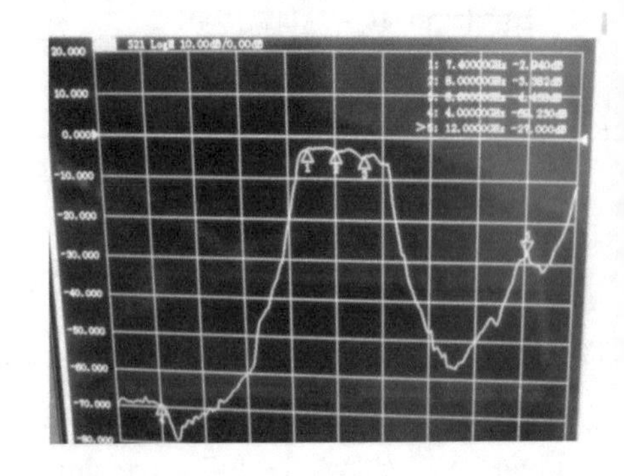

Fig. 5.16 Measured results of the proposed filter

It measures the indicators of the filter by using a vector network analyzer, as show in Fig. 5.16. The test results show that the insertion loss is less than 4 dB, and out of band rejection is about 27 dB, which conforms to the requirements of the indicator, and the filter meets the design requirements.

5.5 Conclusion

This paper studies microwave filters, and designs a C-band interdigital and hairpin bandpass filter, which is used in the transceiver channel. By means of the electromagnetic simulation software, We simulate the interdigital bandpass filter, and use the film substrate to make the filter. The test results show that the insertion loss and out of band rejection meet the design and usage requirements. The microwave passive circuit is an important part of the transceiver module, and affects the performance of the system. Since the microstrip circuit has a simple structure and is easy to integrate, it can be widely used in various microwave circuits.

References

1. Kaczman, D.L., M. Shah, and N. Godambe. 2006. A single-chip tri-band (2100, 1900, 850/800 MHz) WCDMA/HSDPA cellular transceiver. *IEEE Journal of Solid-State Circuits* 41(5): 1122–1132.
2. Levy, R., R.V. Snyder, and G. Matthae. 2002. Design of microwave filter. *IEEE Transactions Microwave Theory Techniques* 50(3): 783–793.
3. Jarry, P., and J. Beneat. 2008. *Advanced design techniques and realizations of microwave and RF filters*. New York: Wiley.
4. Salamat, C.D. 1995. Design of a narrowband hairpin filter on PTEE laminate. *Philippine Engineering Journal.*
5. Lujambio, A., I. Arnedo, M. Chudzik, I. Arregui, T. Lopetegi, and M.A.G. Laso. 2011. Dispersive delay line with effective transmission-type operation in coupled-line technology. *IEEE Microwave and Wireless Components Letters* 21(9): 459–461.

Chapter 6
Research on High-Reliable Low-Loss HTCC Technology Applied in Millimeter Wave SMT Package

Pang Xueman

6.1 Introduction

Recently, with the continually increasing frequency of microwave monolithic integrated circuits (MMIC), the requests for the microwave performance of package are becoming harder. A good microwave package requires small insertion loss and large return loss. Different RF ports need high isolation. With the advantages such as steady dielectric constant and small dielectric loss, Al_2O_3 ceramic has been widely applied in microwave and millimeter wave package research. In addition, HTCC technology can route through several layers and is easier for smaller products. So the HTCC technology based on Al_2O_3 ceramic is widely used in aviation and aerospace communications which have strict requirements about miniaturization and reliability. This chapter will study the millimeter wave package based on HTCC technology.

6.2 Research Work

6.2.1 Low-Loss Ceramic

As a ceramic package material, alumina is widely used because of its high strength and resistance, favorable and low cost [1]. The dielectric loss of alumina ceramic substrate has been researched in industry since it has been used for the microwave applications [2]. Although, the so-called high purity alumina has a low dielectric loss at $\tan\delta = 1 \times 10^{-5}$, the dielectric loss of alumina ceramic substrate is effected by

P. Xueman (✉)
Li Yongbin, Cheng Kai, Wang Ziliang
e-mail: pangxueman@163.com

K. Kuang, R. Sturdivant (eds.), *RF and Microwave Microelectronics Packaging II*,
DOI 10.1007/978-3-319-51697-4_6

the impurity, porosity, microstructure degrade to $\tan\delta = 4 \times 10^{-3}$. That could be tolerated in low frequencies, but fatal limitation in microwave application [3].

Dielectric losses fall into two categories: intrinsic and extrinsic. It is presented that intrinsic losses are dependent on the crystal structure and can be described by the interaction of phonon system with the ac electric field. The ax electric field alters the equilibrium of phonon system and subsequent relaxation is associated with energy dissipation. These intrinsic losses incarnate loss in pure "defect-free" single crystal and set the lower limit of loss in real material [4]. Extrinsic losses are associated with imperfections in crystal structure, e.g., impurities, microstructural defects, grain boundaries, porosity, microcracks, and random crystallite orientation. It is substantiated that the dielectric loss in alumina substrate is limited by these extrinsic factors. It was demonstrated that the impurity and porosity will obviously affect dielectric loss of sintered alumina [5].

Most of high temperature co-fired alumina ceramic substrates used in industry are fabricated by A-95 ceramic, that Al2O3 content ≥95% in the ceramic [2]. Sinter aids, content up to 5%, were added to commercial Al_2O_3 powder to reduce the sintering temperature. Tape casting is the mainstream technology to manufacture thin film which were cut to different precision shape and laminated in a stack to fabricate a substrate. The co-fired process is conducted in humid hydrogen atmosphere to protect printed circuit oxidation. The ceramic sintered in the co-fired process is hard to obtain high density, usually with a porosity of 5%. Concatenate impact of impurity and porosity, dielectric loss of alumina ceramic substrate is as high as 4×10^{-3}. What' s more, sinter aids in liquid phase sintering process form glass phrase in crystalline, which has a high dielectric loss, degrade loss of alumina ceramic substrate.

Commercial Al_2O_3 powder (CL3000FG, ALMATIS, USA) was utilized for fabrication of alumina ceramic substrate.

3 wt% MgO admixed to raw Al_2O_3 powder as Mg-metal salt solution. Consistence and temperature were controlled to force Mg-metal salt deposits on Al_2O_3 powder in hydrate, which loses H_2O in baking. After baking, alumina powder mixed with sinter addition, such as SiO_2, Cr_2O_3, MoO_3, was added in a solvent of alcohol and PVB as binder. Stable slurry with suitable viscosity was obtained after 24 h of ball milling. Such prepared slurry was used for tape casting on a standard tape casting machine after degas for 5 min was casted. The final thickness of the cast tape seals was around 0.20 mm after dried. Different shape samples for different measurements were manufactured by multilayer ceramic technology, involving punching, cutting, and lamination. Samples sintered by two-step sintering: first step, fired at 1600 °C for 60 min and second step, fired at 1650 °C for 60 min in humid hydrogen atmosphere.

Flexural strength was measured by three-point bending method, density by Archimedes method, and porosity by photographic measurement. Samples were characterized by XRD and SEM. The dielectric loss was measured using a high Q cavity resonator method. All dielectric measurements were carried out at room temperature in air at a relative humidity of approximately 30%. The as-fired samples were polished before measuring.

Effect of MgO additive in alumina ceramic was illuminated. As content within solid dissolution limit, 1000 ppm at 1630 °C, MgO dissolves in Al2O3 grain and becomes defective in crystal lattice, which elevate diffusion and facilitate sintering. As content over solid dissolution limit, MgO reacts with Al2O3 to form MgAl2O4 spinel in crystalline, which restrains abnormal grain growth [6]. Meanwhile, MgO content increase transfer rate of pores, which are squeezed out before grain growing, make ceramic more density.

In liquid deposit process, MgO crystallizes with unsaturated function group on Al_2O_3 powder surface as nucleation site in Mg-hydrate. After baking, Mg-hydrate becomes MgO nanoparticles attaching to Al_2O_3 powders. Besides, ensure MgO distribution uniformity, surface energy, and activity of Al_2O_3 powder increase in the liquid deposit process. $MgAl_2O_4$ spinel was observed in Fig. 6.1. XRD of alumina ceramic substrate sample, implying glass phrase forming was ebbed during sintering.

Thus, the porosity decreased with MgO adding by liquid deposit and equiaxial grains in suitable size was obtained, as demonstrated in Fig. 6.2. High density, low porosity, and good flexural strength properties were obtained in this chapter.

The dielectric loss was found to depend strongly on the pore volume with only a small degree of porosity having a very marked effect on the loss. Relation between dielectric loss and porosity could be described as function below:

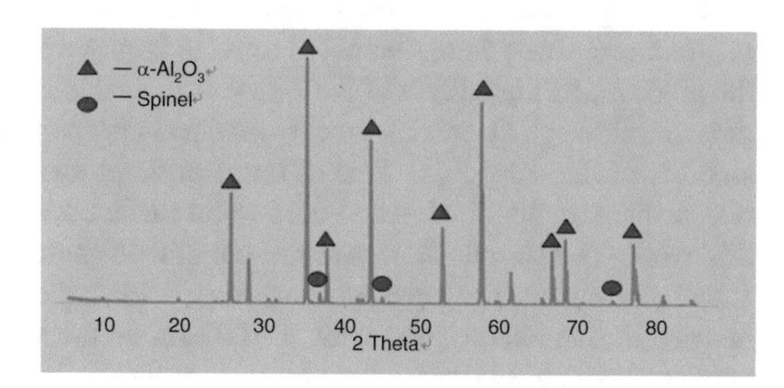

Fig. 6.1 XRD of alumina ceramic substrate

Fig. 6.2 Scanning electron micrograph of alumina ceramic substrate

$$\tan\delta = (1-P)\tan\delta_0 + AP\left(\frac{P}{1-P}\right)^{2/3} \tag{6.1}$$

As porosity decrease for MgO content increasing and liquid deposit process, dielectric loss decreased according to (6.1).

MgO together with sinter additives are added to A12O3 powder in liquid deposit process. So prepared composite powder was shaped by tape casting and sintered by two-step sintering. High density and suitable microstructure were obtained. Given to the decrease of porosity and high loss glass phrase, dielectric property of alumina ceramic substrate was enhanced (Table 6.1).

6.2.2 *Co-fired Tungsten Conductor Pastes*

In order to meet the needs of microelectronic packaging, we selected tungsten powder with two kinds of tungsten powder with globularity and the diameter under 5 μm. Normal quantum of nonmetallic powder and organic carrier were added. It was muller that tungsten powder, nonmetallic powder and organic carrier were blended in together.

The tungsten metallization formula and tungsten slurry production process used in alumina ceramic were studied. In order to improve the stability of tungsten slurry, the process and ingredients of the slurry were analyzed. It was discovered that powders morphology, dispersing process, and resin property had great influences on the stability of the slurry. The time of preparation process, the viscosity of the slurry, reasonable size distribution and the certain surface area of powders were studied in this work. The results show that these methods achieve a very good effect in the actual production. The tungsten slurry is printed on the ceramic substrate which can be used in fine line printing with 50–120 μm on thick film integrated circuits process, as shown in Fig. 6.3.

The THT metallization is a key technique of high temperature co-fired ceramic (HTCC) substrate process, which affects the yield and reliability of final ceramic substrate directly. The corresponding solutions were also introduced. The results displayed that the filling material and tape shrinkage can be controlled to match well

Table 6.1 Dielectric properties of alumina ceramic substrate

Frequency/MHz	ε'	$\tan\delta \times 10^4$
18,783	8.99	10.6
21,743	8.87	10.2
25,294	8.76	9.5
29,197	8.67	9.2
37,467	8.64	8.5

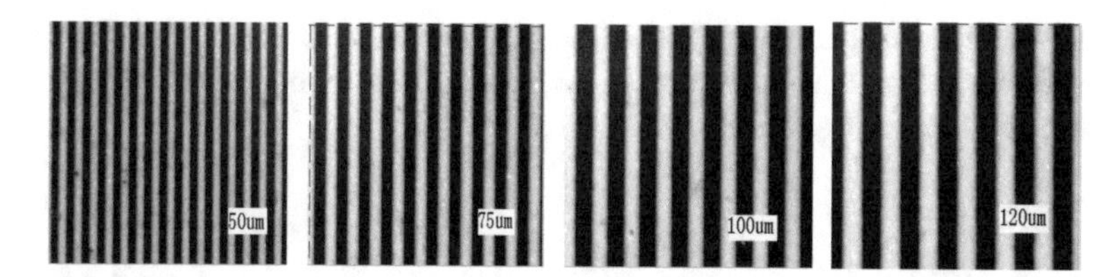

Fig. 6.3 50–120 μm wide printing pattern of tungsten paste

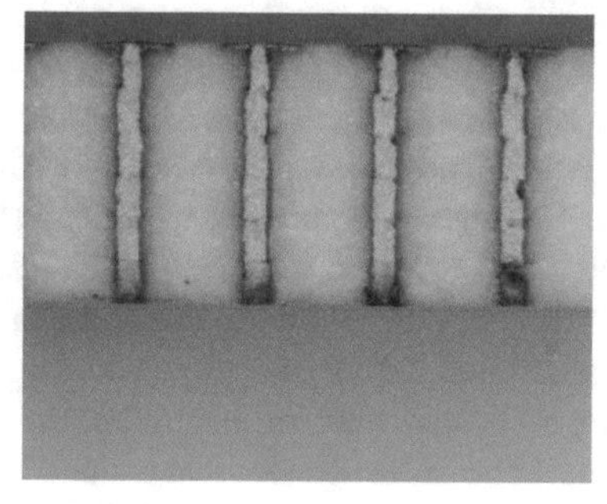

Fig. 6.4 200 μm THT metallization patterns of tungsten paste

by adopting appropriate technical parameter of via-filling process and selecting the filling material with suitable shrinkage and TCE. It is also found that the sintering shrinkage of tape was dominated by the thickness of conductor layer, sintering curve, the temperature and pressure of lamination, and hot-press process. The 200 μm THT metallization patterns are shown in Fig. 6.4.

Low sheet resistance and good matching between substrate and paste are the principal requirements for conductor paste in multilayer co-fired substrate. We need to start their sintering at same temperature and to ensure conductor paste and green-sheet ceramic bodies have the same shrinkage rate.

In this sense, SiO_2 is a good additive. SiO_2 is used as a component in many kinds of pastes. The softening point of pure SiO_2 glass is at about 1600 °C, the same temperature at which greensheet ceramic bodies start shrinkage. SiO_2 can react with Al_2O_3 and form glass. In this chapter, tungsten paste is proposed with SiO_2 glass as the main additive. Complex oxides have low melting point compared to the soaking temperature of Al_2O_3 sintering. The complex oxides become liquid phase in the process of sintering. SiO_2 reacts with CaO, MgO, and $A1_2O_3$, producing CaSiAlO and MgSiAlO phases which are detected at the interface between conductor paste and Al_2O_3 substrate. Casialon glass also forms at the interface. If 0.45% wt SiO_2 is added in the paste, the tungsten paste has low sheet resistance and matching shrinkage of Al_2O_3 greensheet bodies. At this concentration of SiO_2, in the paste, Al_2O_3 substrate is little porous at interface. Al_2O_3 grains at interface are no different from those in the bulk. The sintering stress is decreased to the degree that cannot influence the substrate shrinkage. The size of Al_2O_3 substrate is 50 mm × 50 mm. The resistance of tungsten paste sheet is 6 mΩ/□; metallization combination strength to alumina co-fired ceramic substrate is 54 Mpa.

6.2.3 SMT Package Prepared by the High-Reliable Low-Loss HTCC Technology

The SMT microwave package with 32 pins has been fabricated based on the high-reliable low-loss HTCC technology. The package size is 10.2 mm × 8.2 mm × 0.5 mm only, as shown in Fig. 6.5. The RF signal communication channels use GCPW-VIA-GCPW structure, as shown in Fig. 6.6. The input and output impedances were set as 50 Ω. Through optimizing the influence factors such as line-width, line-spacing of the characteristics impedance of each transmission structure, the insertion loss was lesser than 0.4 dB and return loss was larger than 10 dB in 8-mm frequency range. The simulated result was shown in Fig. 6.7. The result of microwave test on the package showed that the insert loss of single RF signal transmission channel in 8-mm frequency range was lesser than 0.8 dB, the return loss was larger than 10 dB, and the isolation was larger than 20 dB, as shown in Fig. 6.8.

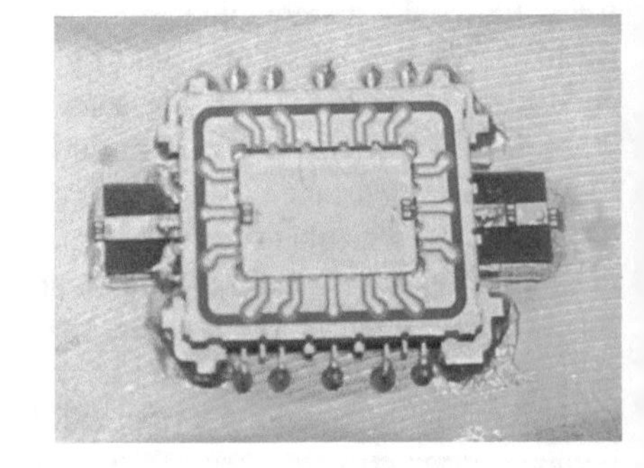

Fig. 6.5 The SMT package aspect

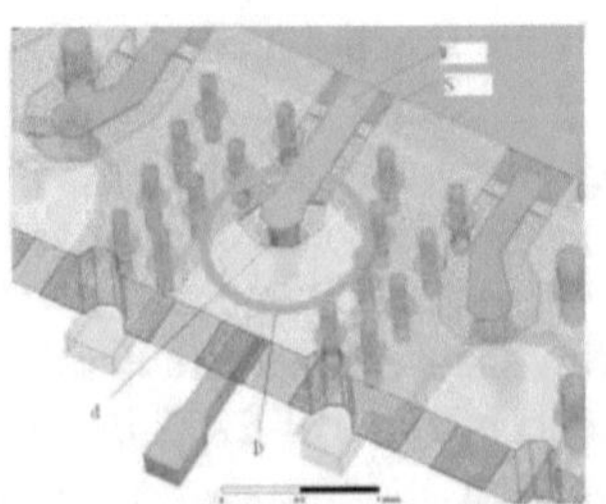

Fig. 6.6 GCPW-VIA-GCPW structure in the SMT package

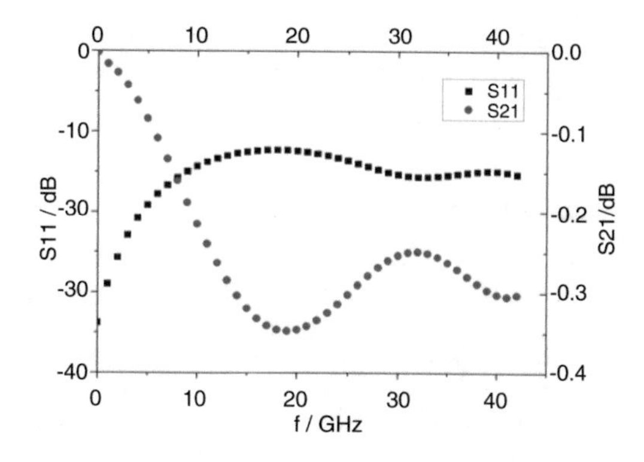

Fig. 6.7 The simulated result of SMT package in this chapter

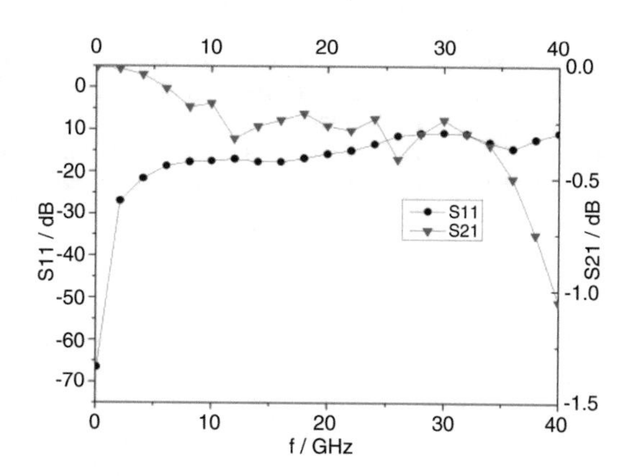

Fig. 6.8 The test result of SMT package in this chapter

6.3 Conclusions

The compositions and technology for high-reliable low-loss HTCC technology were discussed in this chapter. Low-loss ceramic was prepared by Al_2O_3 powder and some additives, MgO, SiO_2, Cr_2O_3, MoO_3, for example. The dielectric loss of the ceramic is $(8.0–10.0)\times10^{-4}$. Techniques of manufacturing high temperature co-fired tungsten conductor pastes were researched in this chapter, and also the effects of microstructure and distribution of tungsten powders, the content of nonmetallic in slurry, metallization combination strength, and sheet resistance of the paste were analyzed. The resistance of tungsten paste sheet is 6 mΩ/□; the metallization combination strength to alumina co-fire ceramic substrate is 54 Mpa. This tungsten slurry can be used in fine line printing of 100 μm line wide on thick film integrated circuits process. The result of microwave test on the SMT package prepared by the high-reliable low-loss HTCC technology showed that the insert loss of single RF signal transmission channel in 8 mm frequency range was lesser than 0.8 dB, the return loss was larger than 10 dB, and the isolation was larger than 20 dB. This research work supported the application of low-loss ceramic in microwave and millimeter-wave medium.

Acknowledgement Thanks are due to analysis center of Southeast University for XRD and SEM analysis. We wish to acknowledge the assistance and support of the RaMP Organizing Committee.

References

1. Swain, M.V. Materials science and technology (vol. 11—structure and properties of ceramics [M].
2. Turnmala, Rao R. 1991. Ceramic and glass-ceramic packaging in the 1990s. *Journal of the American Ceramic Society* 74(5): 895–980.
3. Alford, Neil McN., and Stuart J. Penn. 1996. Sintered alumina with low dielectric loss. *Journal of Applied Physics* 80(10).
4. Gurevich, V.L., and A.K. Tagantsev. 1991. Intrinsic dielectric loss in crystals. *Advances in Physics* 40: 719–767.
5. Perm, Stuart J., Neil McN Alford, Alan Templeton, et al. 1997. Effect of porosity and grain size on the microwave dielectric properties of sintered alumina. *Journal of the American Ceramic Society* 80(7): 1885–1888.
6. Park, C.W., and D.Y. Yoon. 2000. Effects of SiO_2, CaO and MgO additions on the grain growth of alumina. *Journal of the American Ceramic Society* 83(10): 2605–2609.

Chapter 7
Chip Size Packaging (CSP) for RF MEMS Devices

Li Xiao and Honglang Li

7.1 Introduction

Currently, there are two types of CSP packaging techniques: Flip-chip CSP (FC-CSP) [1] and Wafer-level CSP packaging (WLCSP) [2, 3]. FC packaging technology has been in use for over 40 years. It was first introduced by IBM in the 1970s and was subsequently adopted by other chip makers. The main advantage of FC-CSP packaging is its small size. Although there is no definite rule for how small the package should be, a typical FC-CSP packaged device is 50–100% larger than its original die (chip) size. It is a significant improvement from the traditional wire-bonding-based (SMD and QFN) packaging technologies [4].

In contrast, WLCSP is a recent technique that has attracted much attention. Chips packaged using WLCSP are even smaller than the ones packaged using FC-CSP technique. WLCSP also employs a great deal of front-end processes (photo lift, sputtering, etching, etc.). Therefore, many people working on this technology refer to it as a "medium" process (half front-end and half back-end) rather than back-end process.

In the following sections, we will discuss the details of CSP packaging for RF MEMS chips. Although there are many types of RF MEMS devices (switches, filters, tunable capacitors, and phase shifters), micro-acoustic wave filters (BAW and SAW filters) are the most widely used RF MEMS devices at the moment. Hence, the present chapter will focus on the packaging for SAW and BAW filters.

The original version of this chapter was revised. An erratum to this chapter can be found at DOI 10.1007/978-3-319-51697-4_12

L. Xiao
CETC Chongqing Acoustic-Optic-Electronic CO., LTD, Chongqing City, China

H. Li (✉)
Institute of Acoustics, Chinese Academy of Sciences, Beijing, China
e-mail: lhl@mail.ioa.ac.cn

K. Kuang, R. Sturdivant (eds.), *RF and Microwave Microelectronics Packaging II*,
DOI 10.1007/978-3-319-51697-4_7

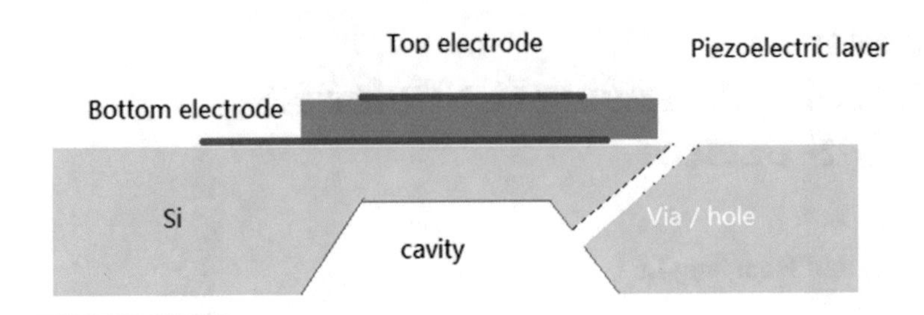

Fig. 7.1 The illustration of a FBAR resonator

7.2 Challenges for RF MEMS Packaging

As mentioned in Sect. 7.1, both FC-CSP and WL-CSP are mature technologies that are widely used in the IC packaging industry. However, it was only recently that such technologies have been applied to MEMS or RF MEMS packaging. So far, many MEMS chips (MEMS microphones, Gyros, pressure sensors, etc.) are still packaged using traditional methods such as surface-mounted device (SMD). This is mainly due to the unique structure associated with MEMS devices.

As seen in Fig. 7.1, a typical FBAR (MEMS) resonator contains a three-dimensional (3D) cavity structure and a resonating layer. The resonator's quality factor (Q) can get degraded if water or dust enters the cavity structure through the releasing hole. The central frequency of the resonator can also change if the top electrode is eroded. Therefore, a well-insulated capsulate must be formed in order to protect such a sensitive structure without introducing water or dust into the packaged device. This presents the greatest challenge for most packaging processes, for example, blade dicing, molding, and bumping, create dust and water.

7.3 Traditional MEMS Packaging Techniques and their Limitations

SMD packaging provides a perfect solution for such problems. In Fig. 7.2, the MEMS chip is placed at the bottom of a ceramic concave (ceramic case) with its resonating layer facing upward. The chip pads are connected to the pads on the ceramic case through wire bonding. A metal plate (top sealing) is soldered on top of the ceramic concave providing a completely sealed capsulate. In applications like military or infrastructure communications, nitrogen gas could be introduced into the sealed capsulate during the sealing process. This technology has been employed by surface acoustic wave (SAW) filter and bulk acoustic wave (BAW) filter manufacturers (Murata, Taiyo and TDK-EPCOS). The main advantage for SMD packaging is its high reliability. Nowadays, most SAW and BAW filter manufacturers still use SMD packaging for industrial or infrastructure communication products. However,

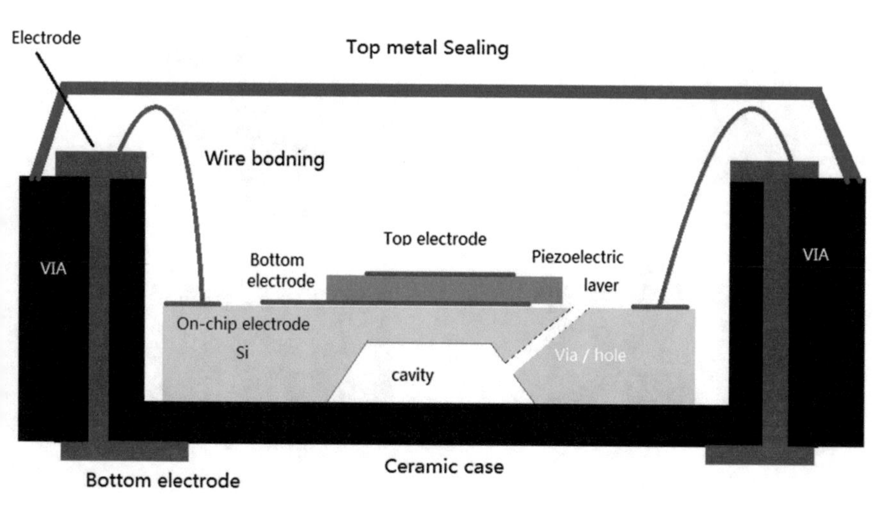

Fig. 7.2 SMD packaged MEMS chips

SMD packaging also exhibits many disadvantages (i.e., large size, costly ceramic case, and slow production rates) making it unpopular for consumer electronics products.

7.4 An Overview of CSP Packaging Technique

CSP packaging is considered a solution to SMD's problems (size, cost, and production rate).

As per our knowledge, there are at least five types of CSP packaging designs depending on how the chip's sensitive 3D structure is protected.

Figure 7.3a shows a classic flip chip package design. The chip is flipped over with its top side containing electronic circuits facing downward to the organic substrate (FR4). A few solder balls or copper pillars (red) are used to connect the in/out signal pads between chip and substrate. Due to the difference in thermal expansion coefficient between solder ball (SnAg) and substrate (FR4), an underfill material must be used to enhance the packaging strength. Another benefit of employing the underfill is to protect the solder balls from corrosion. This type of packaging cannot be applied to MEMS products, because the underfill material would seriously contaminate the MEMS structure (resonating layer).

Although an underfill-free process is technically possible, it is an unusual and highly challenging technique to be used. So far, most original equipment manufacturer (OEM) service providers including Hua Tian Technology and Jcap. Corp (ranked 1 and 2 in the Chinese packaging market) refuse to provide such services. As far as we know, only one SAW/BAW maker, i.e., TDK-EPCOS employs this packaging process.

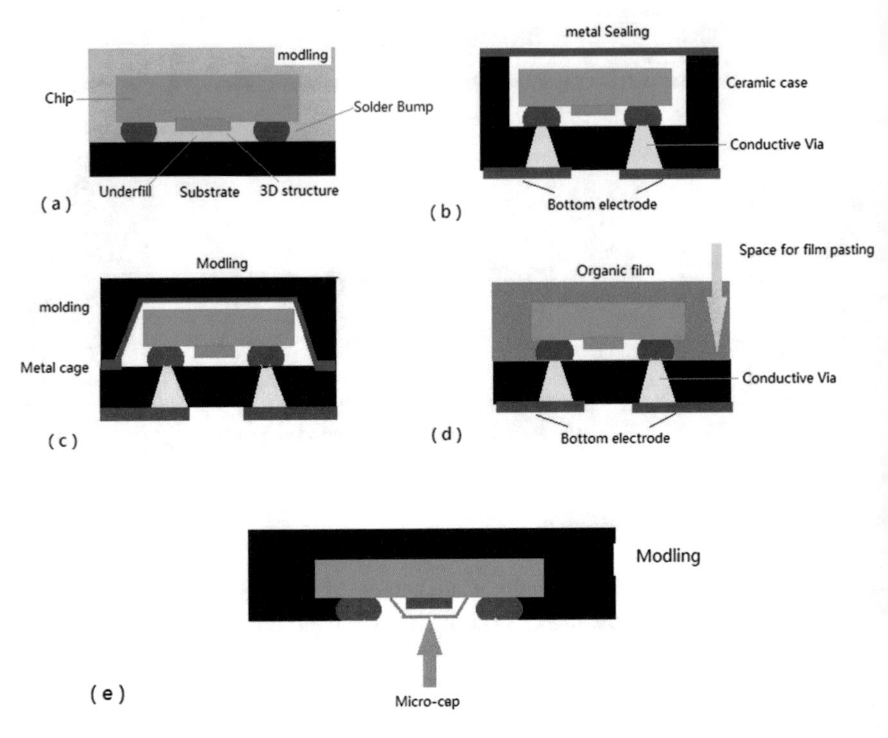

Fig. 7.3 CSP packaging designs

The alternative solution (Fig. 7.3b) is to employ gold stubs instead of solder bumps/balls. Gold is softer and more flexible than a solder ball. Thus, an underfill material is not required for strengthening the flip-chip structure. The ceramic case contains a large central cavity. The signal pads are located at the bottom of the ceramic case. The gold bumps are connected with the external pads though an internal via.

The packaged device shown in Fig. 7.3b is smaller (2.5 mm × 2.0 mm) than SMD-based devices. However, this type of packaging solution is rarely used by any chip maker, because the ceramic case is too expensive for consumer electronic applications. Another reason could be the dwindling supply base for ceramic packages. There are very few suppliers for ceramic cases/substrate (NTK Japan, Korea, and SFE Korea). The average lead time for ordering a customized ceramic package is 2 months! The only hope for using this packaging technique is in areas like aviation, military, or infrastructure communications. One interesting application is the so-called small cells, where Huawei (China) is looking for a replacement for wire-bonding SMD-packaged SAW filters. In small-cell applications, filter reliability is considered a top priority while pricing and size are less important.

The third packaging design (Fig. 7.3c) is more widely used by the industry, especially at the early stage of BAW/SAW development. In this case, the chip is on top of a flat substrate, made of either ceramics or organic (Fr4) material. A metal cap, made by stamping, is used to cover the entire substrate, thus forming a perfectly sealed cavity. Lastly, a molding compound is used to cover the metal capping. This packaging design is much cheaper and smaller than design 3b. The only drawback for this technique is the supplier of miniaturized metal cap. The stamping of small metal cap is a unique process and the IC packaging industry is unfamiliar with it.

The fourth design shown in Fig. 7.3d is perhaps one of the more commonly used packaging technologies for SAW filter. It was initially designed for the packaging of LED chips. In recent years, it has also been used for BAW filter packaging, thanks to the development of stealth laser dicing (detail discussion in Sect. 7.5.2). In Fig. 7.3d, the molding compound is replaced by a special organic film with high viscosity. The film material softens at higher temperatures (60–120 °C) but still exhibits a high degree of viscosity. Therefore, the material would not enter the 3D structures of the MEMS chip. This process is referred to as film pasting and will be discussed in more detail in Sect. 7.5.3. The main advantage of using film instead of metal cap is the low cost. Furthermore, pasting exhibits excellent production rate of 36 wafers/h. Hence, most SAW filter manufacturers have considered employing such process for their products especially in South Korean, China, and Japan [5].

However, organic films provide neither very good sealing against humidity nor EMI (electromagnetic) shielding. This can be a problem for RF MEMS filters used inside a power amplifier (PA) module where the PA is less than 1 mm away from the filter. The RF interference from the PA would significantly affect the performance of SAW/BAW filter. A solution for EM shielding has been reported by TDK-EPCOS (CSSP or CSMP packaging) [6, 7]. The method involves adding a very thin layer of conductive layer on the top or bottom of the organic film to form EMI shielding. This technique also helps improve the sealing of the packaged device. Unfortunately, EPCOS has not published how they add such metal layer on the organic film.

The last solution for RF MEMS packaging is shown in Fig. 7.3e. This is actually a standard WL-CSP packaging process (further discussed in Sect. 7.6). The main difference of WL-CSP in comparison with FC-CSP is that the protection cap (micro cap) [2, 3] only covers the 3D-structure of the MEMS chip instead of the entire chip. There are several advantages of employing such technology over conventional FC-CSP.

1. This is a substrate-free packaging technique allowing the chip to be soldered on the PCB directly.
2. No extra spaces on the edges of the substrate (indicated in Fig. 7.3d) are required for film pasting. As a result, the size of packaged device is nearly the same as the chip size. This is why it is called wafer-level packaging. Both Skyworks and TDK-EPCOS [8] have employed this technology for filter packaging. According to EPCOS, the smallest packaged WCDMA Band1 duplexer is as small as 1.5 mm × 1.1 mm—almost 35% smaller than the smallest FC-CSP packaged duplexer (1.8 mm × 1.4 mm) currently on the market.

It allows solder ball or copper pillar to be used, because the micro-cap is protecting the chip surface in the first place.

The only concern in this technique is how to form a micro-cap directly on the surface of a MEMS chip (further discussed in Sect. 7.6).

7.5 Gold Bump FC-CSP Packaging

As mentioned above, gold bump FC-CSP is the most widely used packaging technology for RF MEMS devices, especially for SAW filters. Hence, the details of FC-CSP will be presented in the following sections. There are six major steps in FC-CSP packaging: (1) gold stub bumping; (2) dicing; (3) flip-chip bonding; (4) film pasting (molding); and (5) singulation.

7.5.1 Ultrasonic Gold (Stub) Bumping

Gold stub bumping has been in use for decades. It is similar to the traditional wire bonding process (Fig. 7.4). Fig. 7.4 shows the bumping process on a piezoelectric wafer (SAW). A typical bumping machine contains six components: (1) the gold

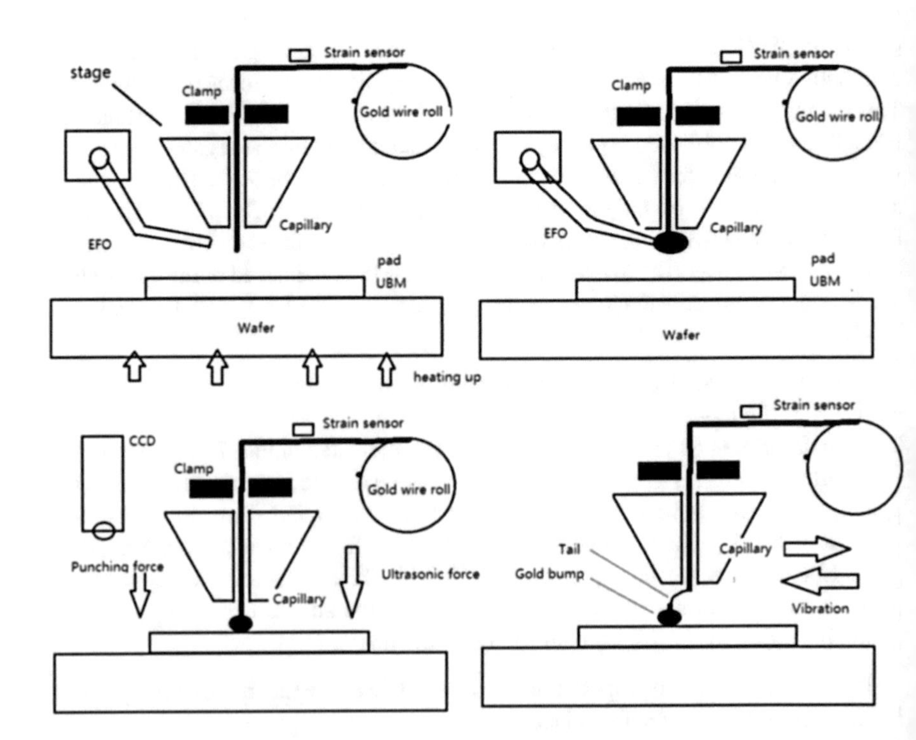

Fig. 7.4 Gold stub bumping steps

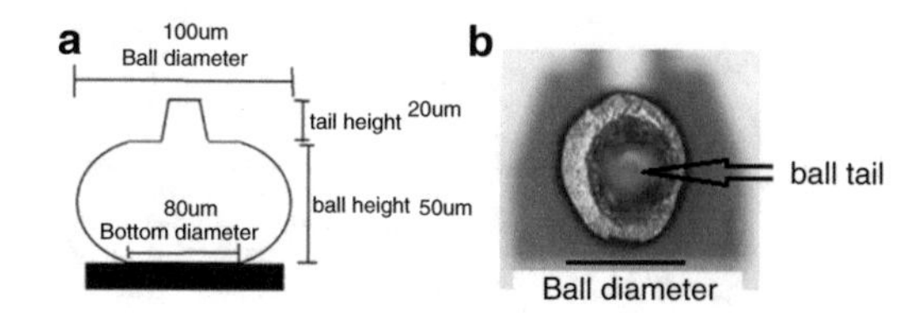

Fig. 7.5 Gold stub formation steps

wire supply chain, (2) capillary, (3) EFO, (4) moving stage, (5) pre-heat stage, and (6) CCD for alignment. First, the gold wire is supplied to the tip of the capillary from the gold wire roll. The wire strain on the roll is constantly monitored by the strain sensor. The clamp controls how quickly the wire is fed through the capillary. The stage would be closed automatically if gold wire breakdown is detected by the strain sensor. Second, the EFO punches the tip of gold wire creating a sudden increase in electrical current. The temperature generated by the electrical current is high enough to melt the tip of gold wire. This allows a gold ball to be formed at the tip of gold wire.

Third, the capillary moves toward the surface of the wafer pads along with the ball-shaped gold wire. Once it touches the pad surface, the ultrasonic generator is turned on (moving sideways) to provide the bumping energy. The punching force provided by the capillary, as it moves to the wafer surface, would make the gold stub wider (more strength). At this stage, the gold stub (ball) is formed on the wafer surface (Fig. 7.5a, b).

Lastly, the capillary moves upward and starts vibration sideways (long distance). The main purpose is to cut off the gold stub tail. The directions of vibration can be in either one or two axes (along X and Y axes). The main concern is to choose a vibration force strong enough to cut the tail off and induce as little vibration as possible on the chip surface. High vibration energy could damage the MEMS structure, especially for SAW filters.

7.5.2 *Dicing*

7.5.2.1 Conventional Blade Dicing for MEMS Chips

Blade dicing is the most widely used method for semiconductor chip dicing. However, the cooling water and dust involved in the dicing process can damage the fragile 3D microstructures of MEMS chips. A simple substitution is to add another carrier (cap) wafer on top of the MEMS wafer before dicing. As shown in Fig. 7.6, the two wafers are temporarily bonded together to prevent any water leakage that may damage the chip cavity. The carrier (cap) wafer can be removed (de-bonded) after dicing by adding a solvent. This process has been used for BAW filter packaging for many years.

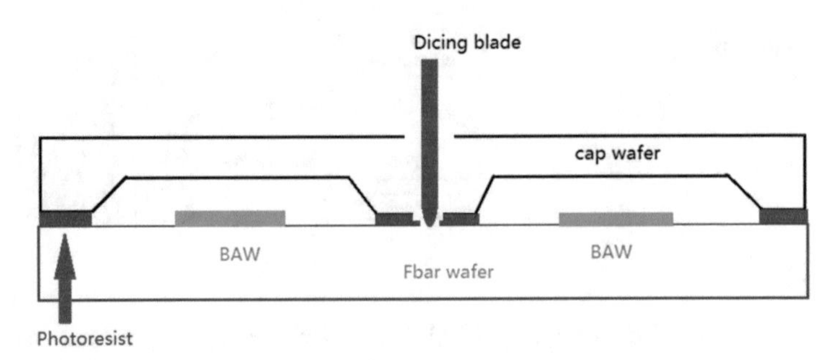

Fig. 7.6 Blade dicing for MEMS wafer with cap wafer protection

7.5.2.2 Stealth Laser Dicing

An alternative to temporary wafer bonding is laser dicing, which is a water-free process. However, laser dicing is notorious for the amount of dust it creates. The laser beam used in dicing would also generate massive heat on the wafer surface. Hence, part of the wafer material is evaporated, especially near the dicing slot. The dust is created by the cooling of the evaporated wafer materials. A simple solution is to place the die very far from each other leaving a wide dicing slot. But this would certainly lower the volume of chips on a single wafer which in turn increases the die cost.

In the past 5 years, another type of laser dicing tool called "stealth laser dicing" [9] has been introduced to the packaging industry. Instead of focusing the laser beam on the wafer surface, the laser beam is focused on the inside of the wafer material. The laser beam weakens the structure inside of the wafer along the dicing slot (Fig. 7.7a). In the end, a so-called "wafer expanding" (Fig. 7.7b) process is used to finally break the bond between each die. In this process, the waked wafer is placed on top of a tape frame. The flat metal stage is moved upward, forcing the tape to be expanded. The tape expansion creates a large stain on the weakened structure, eventually breaking the bond between each die.

This process has been extensively used for the dicing of MEMS microphone wafer. Recently, it has been employed for BAW filter packaging. The only drawback of stealth dicing process is the cost of dicing machinery (one million USD each). However, laser dicing is known for its efficiency. A typical stealth dicing machine (DISCO 7630, Japan) is able to cut a single 6-in. wafer (15 K dies/wafer) in an hour. This has offset the cost of machinery. Another benefit of employing stealth dicing instead of blade dicing is to minimize chipping after dicing. As can be seen in Fig. 7.8, there is barely any chipping that can be observed on the edges of FBAR chip.

7.5.3 Flip Chip Bonding

The step after wafer dicing is flip chip bonding on the substrate. Similar to gold bumping, ultrasonic power has been used for chip bonding. Fig. 7.9 describes the ultrasonic chip bonding process. The chip bonder is made of two main parts: (1) pick up handler and (2) bonding handler.

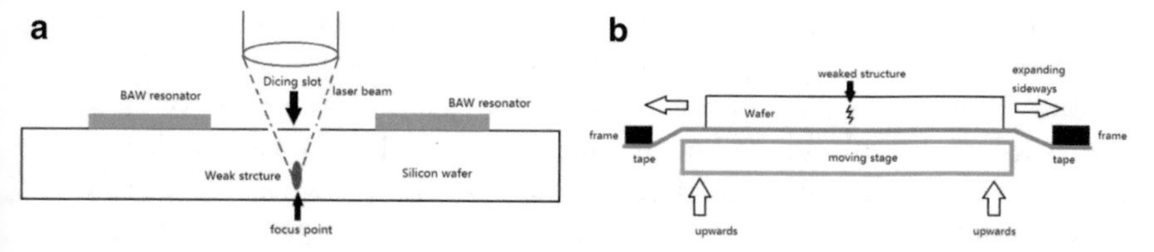

Fig. 7.7 (**a**) Stealth laser dicing; (**b**) expansion process

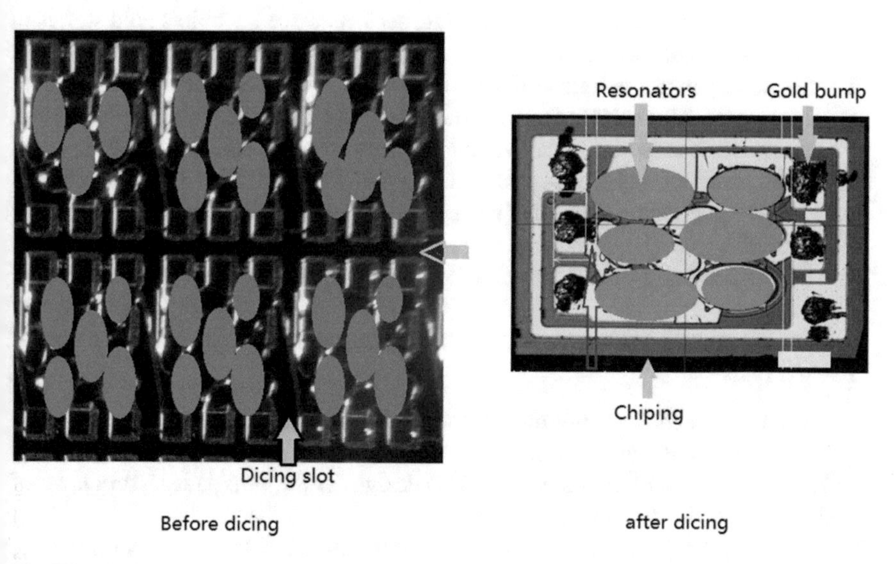

Fig. 7.8 The FBAR chip after stealth dicing

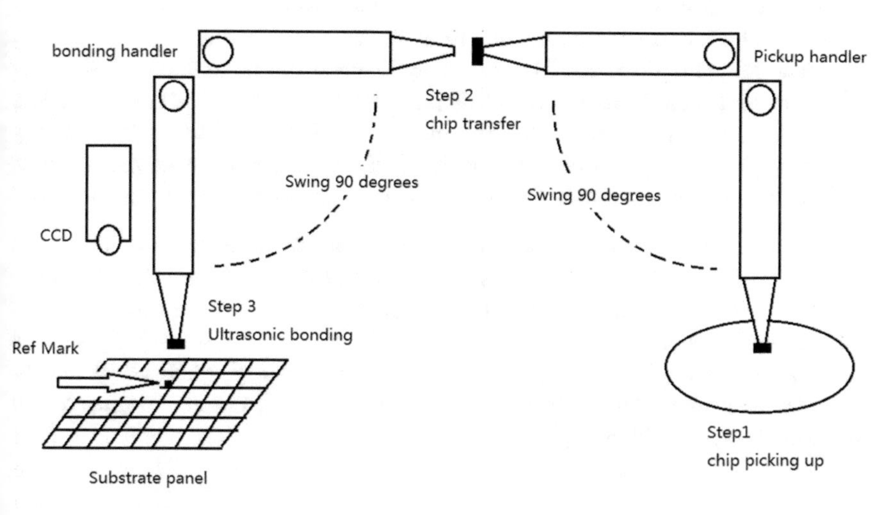

Fig. 7.9 An illustration of ultrasonic bonding process

At first step, the pickup handler picks up the chip from the wafer holder. The tip of the handler is designed with a hollow structure so that the MEMS structure is untouched by the handler. At step 2, the pickup handler swings clockwise by 90° and transfers the chip to the bonding handler. Then the bonding handler swings clockwise by 90° toward the substrate. At step 3, the X–Y stage moves the substrate panel (each panel contains hundreds of individual substrates) to the bottom of the chip. The CCD camera constantly monitors the reference marks on the substrate and ensures that the chip is aligned with the substrate. The CCD camera also carries out visual check for chip integrity (i.e., large chipping caused by dicing). At step 4, the bonding handler moves toward the substrate surface. Once the chip's gold stub is in contact with the pads on the substrate, the ultrasonic power generator is turned on. The ultrasonic energy is transferred to the gold stub and substrate pads. The gold stub instantaneously melts to form a gold joint between the chip and the substrate.

The ultrasonic chip bonding is a mature process, the real concern being the accuracy of the handler precision. A small misalignment of more than 15 μm could seriously affect the bonding strength. Hence, high-accuracy chip bonder with only one or two bonding handlers should be employed to minimize the vibration that affects the bonding accuracy. Hence, the chip bonding rate is not very high (5000 chips/h at best).

There are several advantages for employing gold stub bumping instead of copper or solder balls.

1. There is no need for using underfill to protect the gold stubs which affects the MEMS chip's 3D microstructure.
2. Very simple under ball material (UBM) design for the chip pads. A thin layer of aluminum (200–400 nm) is recommended.
3. It is a very simple process, and thus no extra materials (mask, screen printing, photoresist, and fluxing) are needed.
4. Quick prototyping: Since no extra tooling is needed, gold bumping is extremely popular with smaller chip makers. Gold bumping technique provides a real low-cost but high-performance solution to small-scale prototyping.
5. Small gold stub diameter: The gold stub diameter can be as small as 80 μm if 60 μm wire is used. Gold stub with 80 μm diameter is smaller than most of solder ball (>100 μm) currently provided by the OEM service (i.e., JCAP China). Although smaller gold stub can be achieved, it is at the cost of sheer force strength (<40 N per stub, if the stub diameter is smaller than 60 μm) of the gold bump.

Despite these advantages, gold bumping has gone out of fashion in conventional semiconductor packaging industry, because of the cost of gold wire. Another problem is the slow production rate, where the gold stubs are bumped onto the chip surface one by one. A typical SAW or BAW chip contains six stubs. However, the best gold stub bumping machine can merely achieve a bumping rate of 18 stubs/s. At this rate, the most optimistic estimation is that 240 k chips could be bumped with gold stub (assuming 22 h/day).

7.5.4 *Film Pasting (Molding)*

In case of MEMS chip packaging, a cavity has to be formed in order to protect the 3D microstructure, i.e., BAW resonator and SAW IDT. As mentioned in Sect. 7.4, film pasting is a common way of forming the cavity on the whole chip surface. This technique was especially popular for SAW filter packaging. Recently, it has been employed for FBAR packaging.

As Fig. 7.10a shows, the material used for pasting comes in the form of thick film. The film thickness is between 200 μm and 350 μm depending on the chip thickness. The organic film is manually pasted onto the surface of panel (after flip-chip bonding). Then a flat metal plate with an extra weight is placed on top of the panel to provide the downward force (Fig. 7.10b). The whole structure is inserted into a vacuum chamber to remove any air. Under the vacuum condition, the air gets forced out through the vias/holes of the flat metal panel. The chamber is heated to 60–90 °C to soften the organic film. The organic film still retains a degree of viscosity under high temperature. This allows the organic material to fill all the gaps between each of the die while keeping away from the center area. Fig. 7.11 shows the chip surface view after film pasting.

At last, the panel is heated to 150 °C to harden the organic film by evaporating all the water inside it; this would take about 3 h. There are several advantages of employing film pasting technology. First, it is a highly efficient process where 6–9 panels, each containing 500–800 chips, can be processed simultaneously. The whole softening process takes just under 5 min if a dedicated pasting machine is employed. Second, the film pasting can be easily perfected by adjusting the parameteres such as: vacuum conditions and pressure between the chips and films. As a result, film pasting has been very popular with Asian SAW filter makers for chip packaging.

The last job is panel dicing where the entire panel was cut into hundreds of individual devices. Since the MEMS chips are protected under films and substrate, the dicing speed is very fast and no de-ionized water is required.

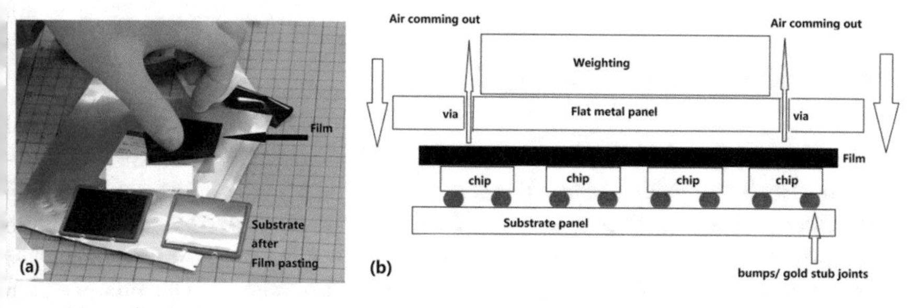

Fig. 7.10 (**a**) Film pasting and (**b**) film softening process

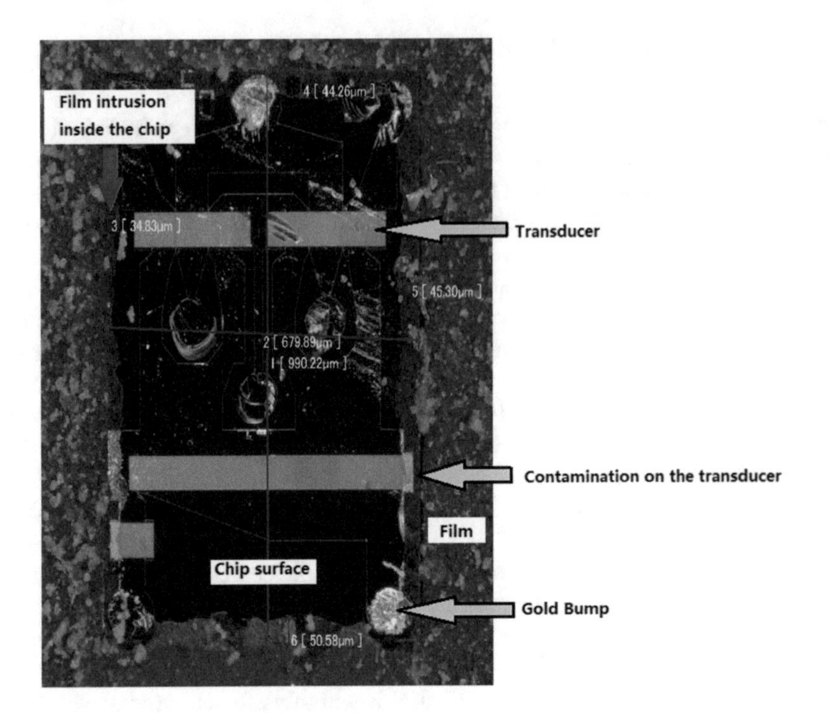

Fig. 7.11 Chip surface view after film pasting

7.6 WL-CSP (Wafer-Level)

An alternative to FC-CSP packaging is the so-called wafer-level CSP packaging (WL-CSP) [2, 3]. As discussed in Sect. 7.1 and Fig. 7.3, the key differences between WL-CSP and FC-CSP is the way the chip's 3D microstructure is sealed. In a typical WL-CSP package, the 3D microstructure is sealed by a microcap or a capping wafer, whereas the entire chip is sealed in an FC-CSP device. Figure 7.12 shows the cross-sectional views of wafer–level-packaged SAW and FBAR filters.

7.6.1 Wafer Level Packaging Design

In Fig. 7.12a, the external pads are connected with the chip pads through a conductive via that is embedded inside the cap wafer. This design is very popular with SAW filter packaging. There are two main reasons for employing such a structure. First, the piezoelectric wafer (LN and LT) used for fabricating the SAW filter is anisotropic, that is nearly impossible to be drilled with a via/hole. The only practical way is to drill via through the silicon cap wafer. Second, the LN and LT wafer sizes (4–6 in.) are much smaller than most of the modern semiconductor wafers (8–12 in.). Hence, most OEM service providers will not provide a 4-in. wafer bonding/packaging or any other micro fabrication service. In Fig. 7.12a, the cap wafers are made

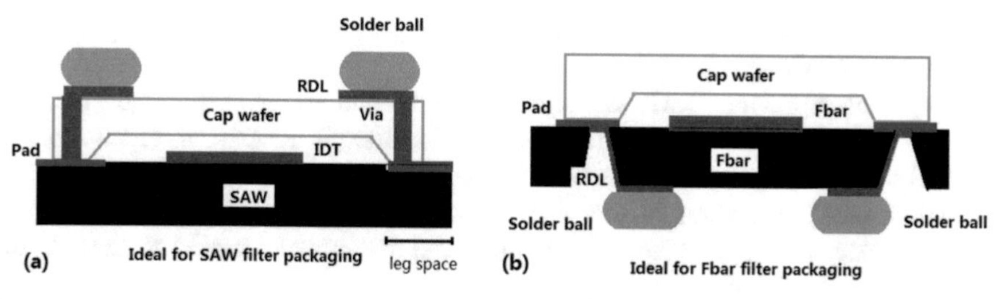

Fig. 7.12 WL-CSP for SAWF and WL-CSP for FBAR filter [10]

from a silicon wafer of size between 8 and 12 in. This allows the foundry (OEM service provider) to handle the packaging process more easily.

The second design in Fig. 7.12b is more popular with FBAR filter packaging industry. As it can be seen, the though holes are embedded into the FBAR chip itself, because FBAR is a silicon based device. In this case, the cap wafer only serves as a protection against surrounding environment.

In both designs (Fig. 7.12), the cap wafers are made from a high resistance material such as glass and high resistance silicon (>10,000 μm/cm). The cap wafer's leg space varies between 50 μm and 400 μm, depending on the application environment (temperature, humidity, and power).

7.6.2 Via/Hole Drilling

The via/hole can be created by laser drilling or deep etching. The laser-drilled holes are then metalized by sputtering. The metal pattern on the cap wafer surface is created by a photolithography process.

7.6.3 RDL Metallization

Both SAW and FBAR filters are operated at GHz frequencies. At these frequencies, the skin effect is the main cause for RF energy loss. The best way of minimizing skin effect is to increase the RDL thickness. Hence, the metal thickness of vias and RDL are further increased by electroplating. Ideally, the RDL metal thickness should be 5–12 μm.

7.6.4 Wafer Bonding

The cap wafer bonding is a well-developed process. However, it is particularly tricky for RF MEMS chips with piezoelectric crystals such as SAW/BAW filters. The piezoelectric wafer (lithium tantalate oxide) has serious warpage issue under high temperature. At 200 °C or above, the LT crystal wafer can break. Hence, most of SAW makers employ an organic adhesive compound for wafer bonding to overcome these challenges.

Organic compounds such as polymers (BCB) and epoxy compounds are frequently used for SAW filters wafer bonding. The organic adhesive material is spin-coated on the cap wafer and patterned to make dams (a wall surrounding the chip's sensitive area) according to the chip layout. The capping chips are (flip-chip) directly placed on the organic compound. The assembly is heated to 200 °C or above to harden the structure [10].

7.7 Conclusion

In this chapter, both FC-CSP and WL-CSP packaging technologies are discussed. Although both technologies are employed by leading players like Skyworks, Murata, and TDK-EPCOS, the WL-CSP packaging is more favorable to the RF MEMS industry because of its small size and low cost. There is also great potential for further improvements in its performance with more accurate wafer bonding machines or laser drilling. In contrast, the gold-bump-based FC-CSP technology has been in use for many years; there is little room left for further optimization. For example, the smallest FC-CSP-packaged SAW duplexer is 1.8 mm × 1.4 mm in size. Further miniaturization on SAW device is impossible with the existing packaging techniques.

References

1. Koch, Robert D., Martin Schwab, and F. Maximilian Pitschi. 2009. Ultra low-profile self-matched SAW duplexer with flip-chip HTCC package for W-CDMA 2100 mobile application. In *International Microwave Symposium Digest*, 97–100.
2. Sakinada, K., A. Moriya, M. Kitajima, and O. Kawachi. 2009. A study of wafer level packaging of SAW filter for module solution. In *IEEE International Ultrasonics Symposium*, 2692–2695.
3. Park, S.W., J.P. Hong, T.H. Kim, S.J. Yang, and J. Ha. 2008. Fabrication and optimization of wafer level SAW filter package using laser via drilling. In *Electronics System-integration Technology Conference*, 1273–1278.
4. Dong, Hao, and T.X. Wu. 2004. Design of miniaturized RF SAW duplexer package. In *IEEE Trans on Ultrasonics, Ferroelectrics and Frequency control*, 849–858.
5. Zhong, J.I.N., and D.U. Xuesong. 2013. HE Xiliang Research of RF-SAW Filter Packaging Technology based on PCB Board, Piezoelectrics & Acoustooptics, 02.
6. TDK-EPCOS. 2012. Smartphones need more than ever—but smaller than ever, http://en.tdk.eu/tdk-en/374108/tech-library/articles/products---technologies/products---technologies/smartphones-need-more-than-ever---but-smaller-than-ever-/172860, 09.
7. TDK-EPCOS. 2013. Trendsetter in packaging technologies, http://en.tdk.eu/tdk-en/373562/tech-library/articles/applications---cases/applications---cases/trendsetter-in-packaging-technologies/763274, 12.

8. TDK-EPCOS. 2011. A new dimension in miniaturization, http://en.tdk.eu/tdk-en/374108/tech-library/articles/products---technologies/products---technologies/a-new-dimension-in-miniaturization/172924, 10.
9. DISCO, Stealth Dicing Application, http://www.disco.co.jp/eg/solution/library/stealth.html.
10. Lau, John H., Chengkuo Lee, C. S. Premachandran, and Yu Aibin. 2010. Advanced MEMS Packaging. Advanced Packaging of RF MEMS Devices, Chapter (McGraw-Hill Professional), Access Engineering.

Chapter 8
The Challenge in Packaging and Assembling the Advanced Power Amplifiers

Cai Liang and Jeff Burger

8.1 Introduction

Transistors have been widely employed in the areas of power amplifiers and power electronics. Packaging and assembly technologies become critical to the success of such applications where high power dissipation (in the form of heat) is involved while operating the transistors. With the ever increasing output power of such applications, heat dissipated from the transistors needs to be more effectively transferred out of the package so that the junction temperature in the transistors can be maintained at a reasonable low level. This is necessary in order to prolong the life time of the transistor devices. Fig. 8.1 shows the predicted device life time versus the junction temperature for typical Si-based transistors as stress tested at 150 °C.

In this example, in order to ensure the device lasts for more than 10 years, the junction temperature must be controlled below 70 °C. Any decrease in the junction temperature will tremendously increase the life time, for instance, if the junction temperature decreases to 60 °C, the life time extends more than 20 years. This plot demonstrates how it is significant that the junction temperature of a device during operation affects the useful life time of a device. This effect on the life time of a device is owning to the change in thermal conductivity, electrical conductivity, impedance and other characteristics of junction with the change in temperature. The primary failure mechanisms of devices in long-term service are strongly dependent on the operating temperature and well modeled using Arrhenius equation as follows:

$$t = A(T) \times \exp\left(\frac{E_a}{kT}\right) \tag{8.1}$$

C. Liang (✉) • J. Burger
Integra Technologies Inc.,
21 Coral Cir, El Segundo, CA90245, USA

K. Kuang, R. Sturdivant (eds.), *RF and Microwave Microelectronics Packaging II*,
DOI 10.1007/978-3-319-51697-4_8

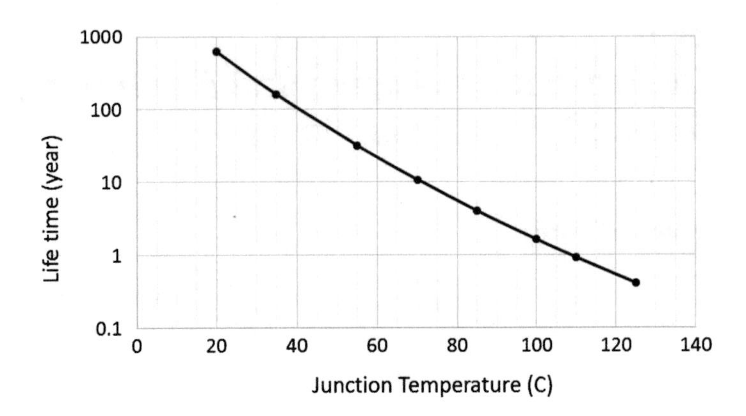

Fig. 8.1 Life time of a transistor vs. the junction temperature

where t is the time to failure, $A(T)$ is a proportional multiplier, E_a is the activation energy in electron volts (eV), k is the Boltzmann's constant, 8.617×10^{-5} (eV/K) and T is the absolute temperature in Kelvin.

Using higher activation energy materials and operating at lower temperature result in longer life time. The well-known outstanding performance of GaAs MMIC has activation energy of 1.6–1.9 eV [1, 2], but its low power density typically makes GaAs non-completive with Si-based LDMOS and BJT in L-band for higher power amplification needs. In recent years, GaN on SiC with activation energy of 0.9–2.2 eV [3, 4] have become an excellent choice as RF transistors for high power amplifier applications due to their higher activation energy with about four times or more power density than GaAs. GaN on SiC also enable higher operating temperature due to the excellent thermal conductivity and high melting temperature of SiC and it is able to perform at high frequency and wide bandwidth. Therefore, LDMOS, BJTs and GaN on SiC are the choice for high power amplifiers in the range of hundreds to thousand Watts. Lowering the operating temperature of a power amplifier module (PAM) [5] is essential: not only can it prolong the transistors' life time, but it also a global benefit to all electronic components on the module since the lower temperature leads to lower creep and fatigue failure rates for packaging and assembly materials. High power density of GaN power amplifiers demands the package substrate and packaging materials possess superior thermal properties. This paper focuses on the challenges that we are facing in manufacturing high power PAMs for RF and microwave applications.

8.2 Thermal Analysis

Heat dissipation is a key concern in PAM devices. A creative geometry design can enhance the dissipated heat to be quickly transferred out of semiconductor chips. Appropriate packaging and assembly materials and techniques not only deliver the heat out of the package faster, but also reduce the impacts of chemical, thermal and mechanical stresses. Figure 8.2 is an illustration of a hermetic version of high power PAM for different applications. Figure 8.2a shows that the transistor is under common base application, where an insulator, or base is placed between the flange and the chips. Figure 8.2b shows that the transistor is under common collector or common source application, where the transistor is directly attached to the package, or flange. Most transistors are fabricated in such a way that the body of the chip is utilized, either as a collector terminal in the case of BTs or as a source terminal in the case of FETs.

The RF input signal also influences the junction or channel temperature of a power amplifier. The signal with continuous waveform (CW) generates more heat than pulsed condition, hence, the saturated temperature in the junction or channel of a PA is higher. Figure 8.3 shows a general junction temperature over time for CW and pulsed conditions.

This is the reason why more attention has to be paid in thermal management of devices when use high power density GaN on SiC for high power PAM application in CW mode.

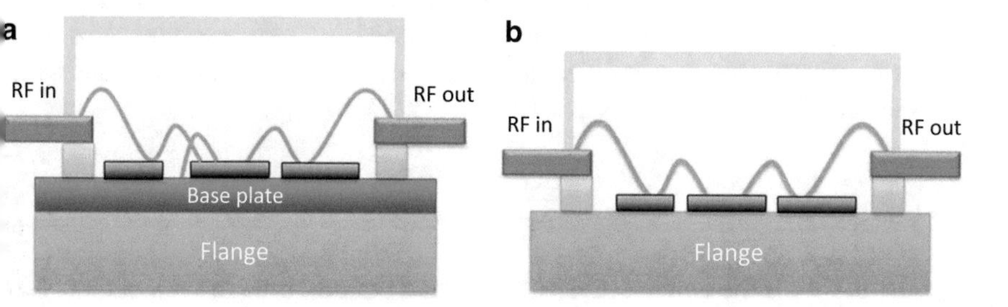

Fig. 8.2 (**a**) Common base application, (**b**) common source, common collector applications

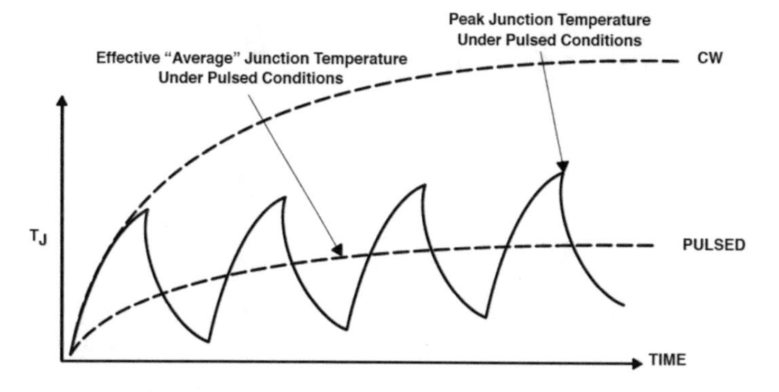

Fig. 8.3 Typical junction temperature change as a function of time development [6]

It is necessary to understand the process of the heat generation and conduction involved in the PAM operation, so that the challenges in PAM manufacturing process from the back end of wafer fabrication to the packaging and assembly processes can be investigated. Figure 8.4 is a mode for a packaged and assembled power amplifier, and its heat conduction analysis. The total increase in temperature from the case or the surface of the heat sink to the junction of the top transistor is expressed in Eq. (8.2).

$$\Delta T = T_j - T_{case} = \sum \Delta T_i \tag{8.2}$$

According to the Fourier's law of thermal conduction, the heat flux density is the amount of energy that flows through a unit area per unit time as described in Eq. (8.3):

$$\underset{q}{\rightarrow} = -k \times \nabla T \tag{8.3}$$

where $\vec{q}$ is the local heat flux density ($W \bullet m^{-2}$); k is the thermal conductivity of a material ($W \bullet m^{-1} \bullet K^{-1}$; and ∇T is the temperature gradient ($K \bullet m^{-1}$).

In the case of PAM, heat generated by the transistor conducts in the thickness direction from top of the junction to the heat sink. If heat loss via radiation is ignored, Fourier's law is simplified in one dimension and in a homogeneous material the heat flow rate can be expressed as:

$$\frac{\Delta Q}{\Delta t} = -kA\frac{\Delta T}{\Delta x} \tag{8.4}$$

where $\frac{\Delta Q}{\Delta t}$ is the amount of heat transferred per unit time in (W), which is the dissipation of power (in term of heat) from the transistor in the PAM case; A is the cross-sectional area, or the surface area perpendicular to the heat conduction direction;

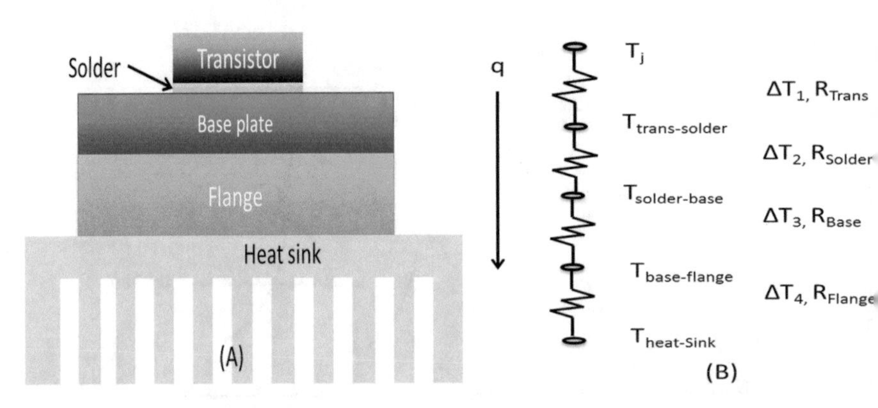

Fig. 8.4. Sketch of power transistor packaging and assemble (**a**), and heat transformation modeling (**b**). ΔT is the temperature difference in the materials, e.g. $\Delta T_1 = T_j - T_{trans-solder}$, T_j represents the temperature on top of the transistor, or the junction temperature, and $T_{trans-solder}$ represents the temperature at interface of the transistor and solder. R is the thermal resistance of a material

ΔT is the temperature difference in two surfaces, and Δx is the distance of two surfaces. Eq. (8.2) can be rewritten as Eq. (8.5),

$$\Delta T = -\left(\sum_{i=1}^{n} \frac{\Delta x_i}{k_i A_i}\right) P_d \tag{8.5}$$

The junction temperature of a transistor can be obtained as

$$T_j = T_c + \theta_{jc} \times P_d \tag{8.6}$$

where $\theta_{jc} = \sum_{i=1}^{n} \frac{\Delta x_i}{k_i \times A_i}$ is the junction-to-case (or to heat sink) thermal resistance (°C/W); T_c is the case (heat sink) temperature; and P_d is the transistor power dissipation. The thermal resistance of a packaged device can be characterized as the sum of a series of resistive components as given below,

$$\theta_{jc} = \theta_{die} + \theta_{die-\text{mount}} + \theta_{\text{base}} + \theta_{\text{flange}} + \theta_{\text{flange-case mount}} \tag{8.7}$$

In order to lower the junction temperature, decreasing P_d and thermal resistance of each series of component is essential for a RF PAM application.

8.3 Challenges Analysis

The heat dissipated by the transistor (P_d) is directly related to the efficiency of a transistor, which is relevant to the design and application. This paper focus on challenge in reducing the overall thermal resistance in PAMs: which are related to the thickness, area, and thermal conductivity of packaging materials and transistors, and their manufacturing process.

8.3.1 Θ_{die}

The thermal resistance of die is affected by die size, die thickness and materials of die itself. Large die size, minimal thickness and high thermal conductivity can reduce the thermal resistance of a die. Die size is determined by the targeting RF application: larger dies size requests larger package size, hence higher packaging material cost is resulted. The choice of die material is only determined by the semiconducting function. The availability is limited to the few types as mentioned before. Reducing the thickness of die is a common method to reduce the thermal resistance. Table 8.1 lists some properties of die and packaging materials including thermal conductivity (TC) and thermal coefficient of expansion (CTE).

Table 8.1 Typical materials' properties [6–10]

Material	Usage	Thickness (×10⁶, m)	Density (g/cm³)	CTE (ppm/K)	T_C (W/m-K) at 25 °C	Specific heat capacity (J/kg-K)
Si	Chip	76	2.31	4.1	150	750
SiC	Chip	76	3.1	4.3	490	690
GaN	Chip	2.5	6.10	5.6	130	490
GaAs	Chip	178	5.3	5.7	50	350
AuSi	Die attach	1	15.7	12.9	190	147
AuSn	Die attach	25	14.7	16	57	15
Au	Die attach, wire bond	2.5	19.3	14.3	310	130
BeO	Substrate	635	2.9	8.0	270	–
W85Cu	Base flange	1000	16.3	7.0	195	–
W80Cu			15.56	9.2	182	188
Mo60Cu	Base flange	1000	9.66	10.3	230	310
CMC (1:2:1)	Base flange	1000	9.54	7.8	260/210[a]	–
CPC (1:4:1)	Base flange	1000	9.46	7.2/9.0[a]	340/300[a]	–
Cu	Base flange	1270	8.95	17	390	380

Note: CMC=Cu:Mo:Cu (1:2:1 thickness ratio); CPC=Cu:Mo70Cu:Cu (1:4:1 thickness ratio)
[a]Donates in the direction of in-plane/through-thickness

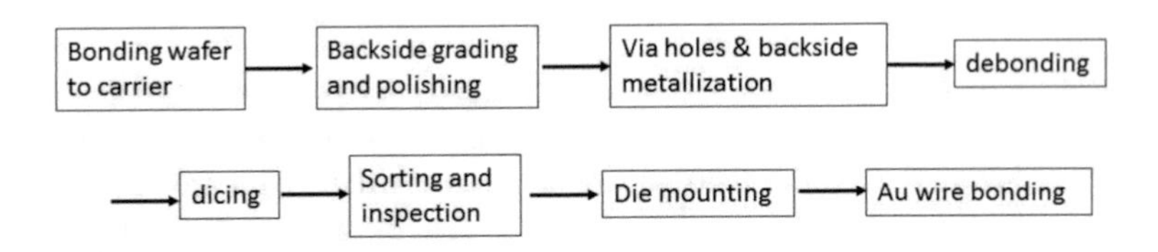

Fig. 8.5 Process flow chart for RF transistor wafer from backside thinning to packaging

Figure 8.5 is a general process flow chart for the back end process after final active IC completion in semiconductor manufacture. In the past decade the commonly used thickness of a transistor die has typically been about 100 μm. Many manufactures have now reduced the thickness down to 50 μm. Reduction in thickness can efficiently reduce the thermal resistance of these lower thermal conductivities semiconductors such as GaAs. Table 8.2 summarizes the thermal resistance changes for a transistor with an area of 5mm², and the thickness reduction from 100 μm to 50 μm.

To achieve this amount of resistance reduction, a significant effort needs to be spent to the processes due to the fragileness of thin wafer: backside grinding and polishing, via holes and backside metallization, wafer dicing, sorting, die-mount and wire bonding.

Figure 8.6 shows a 6″ Si bipolar transistor that was broken during backside processing. High mechanical stress is generally induced during grinding step, even though polishing step may reduce some stress in the wafer. Any initiation of cracks or

Table 8.2 Comparison of thermal resistance change for different transistors with the same thickness reduction

Thickness	Si	GaAs	GaN on SiC	Unit
100 (μm)	0.1333	0.4	0.0408	°C/W
50 (μm)	0.0667	0.2	0.0204	°C/W

Fig. 8.6 A wafer thinned to 2 mils breaks to pieces during process after debonding step

defects on the edge of wafer that are induced by grinding can easily lead to a crack propagating through the whole wafer. The larger the size, the more challenge in handling wafers. In addition, thin die can also be broken during die-mount and wire bond steps. Cleaning of wafer is another challenge for thinned wafer after its de-bonding from its carrier. Large amount of residual can be seen from the photo in Fig. 8.6. The argument is whether it is necessary to thin a GaN-on-SiC wafer down to 50 μm or 75 μm. One can see from the calculation in Table 8.2, the thermal resistance of GaN-on-SiC at 100 μm is even less than a 50 μm thick Si or GaAs transistor, it is 0.0408 °C/W for 100 μm SiC-GaN, 0.0667 °C/W for 50 μm Si and 0.2 °C/W for 50 μm GaAs. Furthermore, backside grinding SiC wafer is even more difficult than Si and GaAs since it is a very hard material. However, it is believed that a thinner SiC may be benefit to fabrications and metallization of back vias, and RF performance as well.

8.3.2 $\Theta_{die\text{-}mount}$

Au-Si is the common solder material for die-attach of Si-based RF power transistors to the package, and Au-Sn is used for GaAs and GaN-on-SiC-based ones. The typical thickness for Au-Si ranges 1–2 μm, which depends on the die-mounting process and the total thickness of Au involved. The thickness of Au-Sn is about 25 μm, depending on the solder preform used. In this case, the thermal resistance is about 0.001 °C/W for Au-Si and 0.088 °C/W for Au-Sn. Using Au-Sn eutectic solder as a die-mount material results in 80 times higher thermal resistance than using Au-Si solder. In addition, the thermal resistance of 1 μm Au-Si solder is about one-fifth of a 50 μm thick Si transistor, which means the Au-Si die-attach material does not impede any heat conduction for Si-based transistor. While the use of a 25 μm Au-Sn solder does improve the thermal conduction for a GaAs, it impedes the thermal conduction of GaN-on-SiC transistor even at a 100 μm thick. Therefore, the challenge in die-mount is how to improve the thermal conductivity of die-mount material by utilizing an advanced material even though Au-Si and Au-Sn have demonstrated excellent thermal fatigue performance. In the later section, we discuss the potential of use nano-silver-particles for high power RF PAM packaging.

8.3.3 Θ_{base} and Θ_{flange}

The base plate is used as an isolation material that is brazed to the flange in a package when a transistor, e.g. bipolar, is used in a common base configuration. It also service as a heat spreader. Beryllium oxide (BeO) is the most commonly used isolation and heat spreader material in the bipolar transistor packaging owning to its higher thermal conductivity and low CTE. In the cases of common collector and common source configurations, typically transistors are directly mounted to a package or flange that is made of W-Cu or Mo-Cu material. This direct die attaching process removes the thermal resistance incurred by the BeO. The flange serves as a die-attach substrate, heat spreader and heat/electrical conductor as simply shown in Fig. 8.7. There are two extreme conditions considered for the heat transferring out of the package. One is that the heat only conducts through the area well below the die-package interface called mode (a), the other is the heat equally spread from the die-package interface out to the package area called mode (b).

The thermal resistances of different packages under the different thermal conduction mode (a) and (b) are listed in Table 8.3. The data demonstrate that, in the worst case when the heat is dissipated only via the die attaching area, the thermal resistance is high,

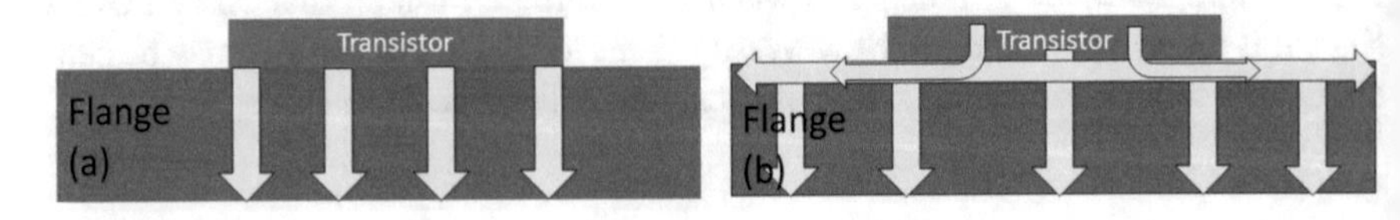

Fig. 8.7 Modeling heat conducting via die contact area (**a**), and via whole flange area (**b**)

Table 8.3 List of total thermal resistances of base and flange for different transistor applications

		Θ_{base} (°C/W)		Θ_{flange}(°C/W)		Θ_{total}(°C/W)	
Transistor type	Package type	Mode (a)	Mode (b)	Mode (a)	Mode 1(b)	Mode (a)	Mode (b)
Bipolar	BeO+W85Cu	0.4704	0.003	1.0256	0.0032	1.496	0.0062
LDMOS	CPC	0	0	0.6667	0.0018	0.6667	0.0018
SiC-GaN	W85Cu	0	0	1.0256	0.0435	1.0256	0.0436

and it is in the range of 0.68–1.5 °C/W, but it is very small in the perfect case when the heat is dissipated evenly through the package body down to heat sink. In reality the thermal resistance value falls between these two extremes. The challenge here is how to select the best package material so that the heat can be quickly spread out both in the plan direction and in the thickness direction. Copper has been long time used for this purpose but it was replaced by W-Cu, Mo-Cu, CMC or CPC to reduce the CTE between die and package while sacrificing some thermal conductivity. In the case of pulse input signals, for low duty factors of 1%–2%, the power dissipation from a transistor is very small. However, when the input signal is in continuous waveform, the dissipated heat from transistor is 50 to 100 times higher; even the small thermal resistance results in a large temperature gradient, hence the junction temperature rises significantly.

8.3.4 $\Theta_{flange\text{-}case\ mount}$

The RF PAM package is typically mount to a PCB (system board) with the flange directly mount to heat sink or case by solder material or bolt down attachment with thermal pads, which introduces about 0.001–0.01 °C/W thermal resistance in our case. Solder mounting techniques provide the best thermal conduction than other techniques. The temperature of the case or the heat sink directly influences the junction temperature of the transistor.

8.4 Case Study

Three types of transistors, named bipolar, LDMOS and GaN on SiC, are used as examples to analyze the transistor's junction temperature for different applications. In order to simplify the study, only silicon is considered as the transistor construction material in bipolar and LDMOS transistors, and only SiC is considered in GaN on SiC. The thermal resistance at the interface of BeO and flange is ignored. Assuming that the transistor has a thickness of 100 μm and an area of 5 mm^2, the thermal resistance of each component is listed in Table 8.4 for different transistors at different heat conducting modes. The thermal resistance of die attachment is considerably high compared with the die itself. The thermal resistance for mode (a) is much higher than mode (b) due to small conductive area, and CPC results in less resistance due to its high thermal conductivity.

Based on the thermal resistance of different types of transistor in Table 8.4, Table 8.5 is the comparison of the junction (or channel) temperature of transistors with pulse and CW inputs. The maximum and minimum junction temperatures are obtained using the model described in Fig. 8.7a and b for different RF PA applications, respectively. It shows that the current design is good enough for pulse input application. However, if these RF amplifiers operate at CW mode, the temperature at the junction ranges from 160 °C to 570 °C when the heat is only conducted through

Table 8.4 Breakdown thermal resistance for different types of power amplifiers

Transistor type	Package type	Θ_{die} (°C/W)	$\Theta_{die\text{-}mount}$ (°C/W)	Θ_{base} (°C/W)		Θ_{flange} (°C/W)		Θ_{total} (°C/W)	
		Mode (a)	Mode (a)	Mode (a)	Mode (b)	Mode (a)	Mode (b)	Mode (a)	Mode (b)
Bipolar	BeO+W85Cu	0.1333	0.0011	0.4704	0.003	1.0256	0.0032	1.6304	0.1406
LDMOS	CPC	0.1333	0.0877	0	0	0.6667	0.0018	0.8877	0.2228
SiC-GaN	W85Cu	0.0408	0.0877	0	0	1.0256	0.0435	1.1541	0.172

Table 8.5 Comparison of junction temperature (T_j) under different operation modes for different transistors

Transistor type	Case temperature (°C)	Θ_{total} (°C/W) Mode (a)	Θ_{total} (°C/W) Mode (b)	P_d (W) Pulse	P_d (W) CW	T_j (pulse) (°C) Mode (a)	T_j (pulse) (°C) Mode (b)	T_j (CW) (°C) Mode (a)	T_j (CW) (°C) Mode (b)
Bipolar	55	1.6304	0.1406	3.14	314	60.1	55.4	566.9	104.5
LDMOS	55	0.8877	0.2228	2.38	119	57.1	55.5	160.6	81.5
SiC-GaN	55	1.1541	0.172	4.02	201	59.6	55.7	287.0	89.6

the area of die size as described in Fig. 8.a7. Under such extremely high temperature devices cannot function properly. On the other hand, when heat conduction is assumed in the perfect mode as described in Fig. 8.7b, the junction temperatures can be lowered down to the range of 80 °C–105 °C with respect to the case or heat sink temperature of 55 °C. In fact, the junction temperature is much higher than this value, since the heat cannot be effectively spread out in the plane of flange or package. High junction temperature can significantly de-rate the life time of device even though it is still under the rating temperature. In the case of CW operation, the current thermal management design is not applicable to bipolar RF amplifiers, and significant improvements for LDMOS and GaN are essential. Figure 8.8 shows the junction temperature of RF amplifier that operates at CW mode changes with the ratio of conduction, where the ratio is the conducting size to the transistor size. From Eq. (8.5) the temperature change is the reverse to the conduction area, but when the heat conduction area is large enough, the further drop in temperature is limited. In additional, large area results in large size and high cost. Hence, the use of higher thermal conductive package material is necessary to compromise the package size.

8.5 New Approaches

Figure 8.8 shows the reduction of junction temperature is limited by increasing the conduction area through the package when the area is large enough. Forced cooling techniques can be used to lower the case/heat sink temperature to room temperature to maintain a low case temperature. However this is constrained by many factors in a subsystem for the end users. Once the semiconductor material is defined, the selections of die-mounting and package materials are possibly the only options for the thermal management design for the RF PAM.

8.5.1 Package Materials

In order to meet the request for high power RF PAM, many companies have actively worked on different approaches to reduce the thermal resistance. Momentive and SMI nano Materials developed an embedding technology using graphene and

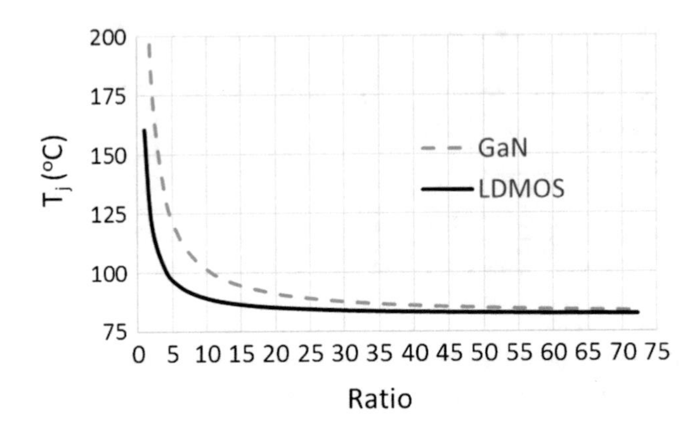

Fig. 8.8 Junction temperature changes as a function of the ratio of heat conducting through the area of flange. The ratio is the conducting area to the transistor size

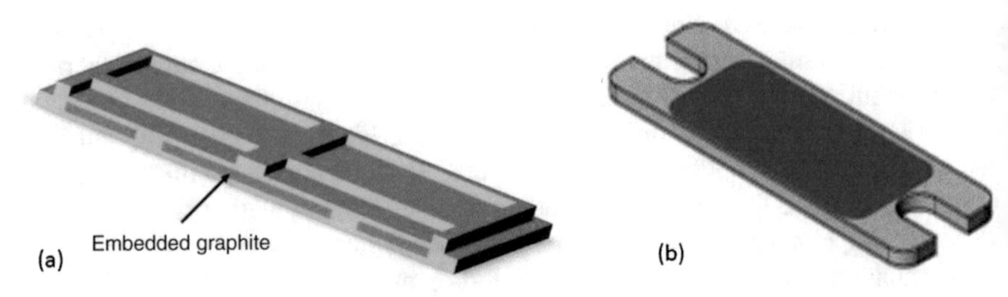

Fig. 8.9 Sketches of embedded graphite or Al-diamond heat spreader in a package: (**a**) embedded in a partial of package, (**b**) embedded in full area [11]

aluminum-diamond as embedding materials inside the package or flange as shown in Fig. 8.9, such that the heat can be efficiently spread out laterally and transferred out vertically from the package. Both graphene and aluminum-diamond have thermal conductivities about 500 W/m-K or higher. Moreover, their CTEs are around 6.5 W/m-K, which makes them very compatible with semiconductors. The cladding materials must have high TC and low CTE to match the CTEs of die and the embedded heat spreader. Sumitomo is also in the process of developing Ag-diamond composite as heat spreader that is to be bonded to the flange without use of cladding material on top of it.

8.5.2 *Die-Mounting Material*

The use of 25 μm Au-Sn eutectic solder for die-mount induces 0.088 °C/W thermal resistance for an area of 5 mm^2 die, which is twice the value of 100 μm thick GaN on SiC chips or four times of the value for 50 μm thick chips as in Table 8.2.

The low thermal conductivity of AuSn is one of factors that hinders the reduction of junction temperature. In recent years, silver nano particles have been under development by many companies [12–14]. The silver nano particle sintering material is different from the silver paste that is also used for low power die attach applications. The conventional silver paste is a polymer-based bonding material with a small percentage of silver loading and the bonding occurs during the polymer cross-linking process. Conversely, the nano silver particle material is a silver-based material with organic components: the bonding takes place during the sintering of nano particles. Nano silver particles can be solidified at a relatively low temperature. After sintering it becomes a solid form silver with the similar thermal and mechanical properties as its bulk counterpart. The advantages of using sliver particles as a die-mount material to replace solders are illustrated as in the following:

1. Thermal conductivity of Ag nano particles after sintering can be up to 250 W/m-K or more, which is higher than both Au-Sn and Au-Si solders, hence resulting in better thermal management.
2. The nano silver particles can be sintered at a low temperature as low as 200 °C, which can greatly reduce packaging and assembly process temperature, hence the stress induced in the transistor during manufacture process can be minimized.
3. The unique properties of nano particles can reduce the die-mounting steps to a single sintering step. Traditional die-mount involves 410 °C for Au-Si and 310 °C for Au-Sn two steps, hence the improvement of productivity is possible.

8.6 Conclusions

Thermal management is crucial to high power RF amplifier working under continues waveform type conditions where the dissipation power is high. To achieve a lower operating temperature on the transistors, challenges involved in the manufacture include wafer thinning, handling and packaging processes, the die-mounting materials and the package/flange materials selection. Each thermal resistance component in a typical RF power amplifier is analyzed. Reduction of the thickness of GaN-on-SiC to attempt lower thermal resistance incurred in the die is not appreciated unless it is requested from the RF performance needs. Any small decrease in the thermal resistance in the die by reducing the thickness results in high cost of wafer backside thinning and high risk of wafer breakage in the manufacture practices. Using AuSn eutectic solder in die-mount results in a relatively large amount of thermal resistance due to its low heat conductivity and large thickness. BeO properly is the best selection for bipolar transistor packaging materials due to its great CTE and thermal conductivity. However, there are large challenges in the thermal management of continuous waveform application using bipolar transistor for high power RF amplifiers. Case studies demonstrate that the current designs for pulse application are good, but not good enough for CW mode operation. The junction temperatures are about 45 °C and 50 °C higher than the case temperature for RF PAM using GaN and bipolar transistor, respectively. Newly developed nano silver particles for die-mount can potentially reduce the thermal resistance incurred by the AuSn solder, while also reducing the overall thermal stress resulting from the packaging and assembly steps. Moreover, the manufacture time for packaging and assembly can be reduced. Use of highly conductive graphene or metal-diamond composites as heat spreading materials with their incorporation into the package may be one of the thermal management solutions to the high power RF PAM.

References

1. Leung, D. L., Y. C. Chou, C. S. Wu, R. Kono, J. Scarpulla, R. Lai, M. Hoppe, and D. C. Streit. High reliability non-hermetic 0.15 um GaAs pseudomorphic HEMT MMIC amplifiers.
2. GaAs MMIC reliability assurance guideline for space applications; Sammy Kayali, Jet Propulsion Laboratory; George Ponchak, NASA Lewis Research Center; Roland Shaw Shason Microwave Corporation.
3. Wu, Y., C.-Y. Chen, and J. A. del Alamo. Temperature-accelerated degradation of GaN HEMTs under high-power stress: activation energy of drain-current degradation.
4. GCS Qualification report, internal document
5. JEDEC 237, Reliability qualification of power amplifier modules
6. Thermal considerations for RF power amplifier devices, TI
7. http://www.semiconductors.co.uk/
8. Product data sheet-eutectic Au-Tin solder; indium Corporation.
9. http://www.engineersedge.com/properties_of_metals.htm
10. www.torreyhillstech.com
11. 3D graphene for next generation thermal management, Ramesh Varma, Wes Nusbaum, Martin Goetz, and Wei Fan RaMP2016, April 5–6, 2016. San Diego, CA.
12. Buttay, C., A. Masson, J. Li, M. Johnson, M. Lazar, C. Raynaud, and H. Morel. 2011. Die attach of power devices using silver sintering-bonding process optimization and characterization. In *IMAPS*, 1–7. Oxford: High Temperature Electronics Network (HiTEN).
13. Kraft, S., S. Zischler, N. Tham, and A. Schletz. 2013. Properties of a novel silver sintering die attach material for high temperature—high lifetime applications. AMA Conferences 2013—SENSOR 2013, OPTO 2013, IRS^2 2013.
14. Silver sintering paste CT2700R7S, Technical Data Sheet (EN) ver.3, 2016, Kyocera.

Chapter 9
High Thermal Conductivity Materials: Aluminum Diamond, Aluminum Silicon Carbide, and Copper Diamond

Kevin Loutfy, Birol Sonuparlak, and Raouf Loutfy

9.1 Materials Used in Microwave Packages and Expectations from Materials, Availability and Properties of These Materials

Thermal management of RF amplifier devices is becoming more critical as GaN on Silicon Carbide (SiC) and GaN on Diamond devices are being produced which have high power densities that are substantially greater than previous Silicon-based devices.

The management of heat dissipation from GaN on SiC or GaN on Diamond devices to the appropriate heat spreader material presents challenges that existing materials such as Copper Tungsten, Copper Moly, and Copper Moly Copper fail to fully support and as a result these materials are limiting the performance benefits inherent to GaN.

These bottlenecks include temperature de-rating of the device, power de-rating, impact on efficiency, and reliability reduction due to the inability of these materials to dissipate heat.

Heat spreader materials present a unique challenge in that they need a CTE close to the semiconductor material as well as other components in an electronic package (e.g., 4–8 ppm/K) and also have a high thermal conductivity [1]. Existing materials achieve low CTE at the trade-off of thermal conductivity. Most commonly used heat spreaders materials are listed in Table 9.1.

K. Loutfy (✉) • R. Loutfy
Nano Materials International Corporation, 1230 E Speedway Boulevard, #211, Tucson, AZ 85721, USA
e-mail: kloutfy@nanomaterials-intl.com; xiaoli@sipat.com

B. Sonuparlak
High Tech Materials Source, Chandler, AZ, USA

K. Kuang, R. Sturdivant (eds.), *RF and Microwave Microelectronics Packaging II*, DOI 10.1007/978-3-319-51697-4_9

Table 9.1 Material properties of heat spreader materials [1, 2]

Material	Composition	Thermal Conductivity (W/m K)	CTE (ppm/K)
Cu	Pure	393	17.0
Al	Pure	204	23.0
Diamond		1500	1.4
Aluminum diamond		500–550	7.0
Copper diamond		500–550	6.0–6.5
Silver diamond		560–600	6.0–6.2
SiC	4H–Si	430	4.0
AlSiC	63% SiC	>175	7.9
W90Cu	90% W	185	6.5
W75Cu	75% W	225	9.0
Mo70Cu	70% Mo	185	9.1
Mo50Cu	50% Mo	250	11.5
CuMoCu	1:4:1	220	6.0
CuMoCu	1:1:1	310	8.8
Cu/Mo70Cu/Cu	1:4:1 laminate	240	8.0

Based on their thermal conductivity, these heat spreader materials can be divided into two groups. The first group of materials has thermal conductivity below 250 W/m K and the second group has thermal conductivity over 250 W/m K.

Particulate molybdenum (Mo) and particulate tungsten (W) reinforced copper (Cu) and a composite material composed of an aluminum metal matrix reinforced by silicon carbide particles are in the first group (Group 1) and mostly used as traditional high thermal conductivity and tailored CTE materials. The second group (Group 2) of heat spreader materials includes laminated Cu/Mo/Cu (CMC), Cu/CuMo/Cu, and diamond particulate reinforced metal matrix composites where the metal is aluminum, copper or silver.

The thermal conductivity of the first heat spreader material group usually varies between 175 W/m K and 250 W/mK. CTE of W-Cu and Mo-Cu can be tailored between 6.5 ppm/K and 11.5 ppm/K. CTE is tailored by varying amount of W or Mo particle loading. In the desired CTE range (4–8 ppm/K), thermal conductivity of W-Cu and Mo-Cu are usually below 200 W/mK. SiC reinforced aluminum is another first group heat spreader material that has emerged in the early 1980s. AlSiC became the material of choice in many new system designs where light weight, high thermal conductivity, tailored coefficient of thermal expansion, and high stiffness are key design parameters. AlSiC thermal conductivity can be as high as 230 W/mK and CTE as low as 5.5 ppm/K. Most widely used AlSiC composite electronic packages have CTE between 7–8 ppm/K and thermal conductivity between 170–200 W/mK. AlSiC can also be cast to net shape. The superior combination of properties specific to AlSiC electronic packages, combined with net shape casting capability yield performance advantages over competing electronic material systems. Similar to Cu/W, Cu/Mo, and CMC, AlSiC allows for direct component attachment, but is much lighter than those mentioned materials. Users typically achieve a 60% reduction in module size and/or weight through the implementation of an AlSiC material solution. Other widely used electronic package materials are Kovar, titanium (Ti), and other low CTE alloys. These low CTE materials are not classified as heat spreaders because of their low thermal conductivity. Thermal conductivity of these materials is at least 10× worse thermal conductor than AlSiC, Mo-Cu, and W-Cu.

Other high thermal conductivity materials, such as aluminum and copper are not considered as preferred heat spreader material in high-performance demanding applications, because they have performance limitations affecting both reliability and heat dissipation capability. Both aluminum and copper have high CTE (23 ppm/K and 17 ppm/K, respectively) and require low CTE carriers for component attach and/or manufacturing process to compensate for the CTE mismatch between the mounting surface and the electronic components.

Various versions of CMCs included in the second group (TC >250 W/mK) are produced by laminating Cu and Mo and/or CuMo sheets. Thermal conductivity of some versions of these laminated sheets can be lower than 250 W/m-K, especially when thickness of the Mo layer is increased to lower its CTE. These laminated composites replace the CuW composites when higher thermal conductivity is a design requirement. Since these materials are produced by laminating individual Cu and Mo layers, there are challenges in producing flat laminated sheets, plates with strong mechanical bond between layers, and parts that do not have CTE hysteresis. Thermal conductivity of these composites is reported as high as 340 W/mK by increasing copper layer thickness which also increases CTE of the heat spreader over 8 ppm/K. When higher thermal conductivity is required in the CTE range of 4–8 ppm/K, the use of diamond particle reinforced metal matrix composites is the only choice.

Development of metal matrix composite (MMC) materials with tailored CTE (4–8 ppm/K) and thermal conductivity (>450 W/m K), specifically those based on diamond particles (with thermal conductivity between 1000 and 2000 W/mK), has been continued since the early 1980s and commercialization has been watched with great interest over the years as a solution to existing thermal material limitations.

The commercial adoption of diamond-based MMCs has been limited due to cost, ability of the manufacturing base to produce volume quantities, and the ability to meet all the technical requirements necessary for the material to be truly adopted by the customer base.

This chapter explains material developments that have been taken in addressing these issues and providing a high performance thermal management material that can be integrated into devices.

9.2 Role of Aluminum Diamond and AlSiC in Advancing Microwave Technology

One of the most common heat spreader applications is the RF power amplifier. As illustrated in Fig. 9.1, an RF power amplifier is first built by joining different materials including heat spreader, power chip (GaN or GaAs), ceramic ring frames

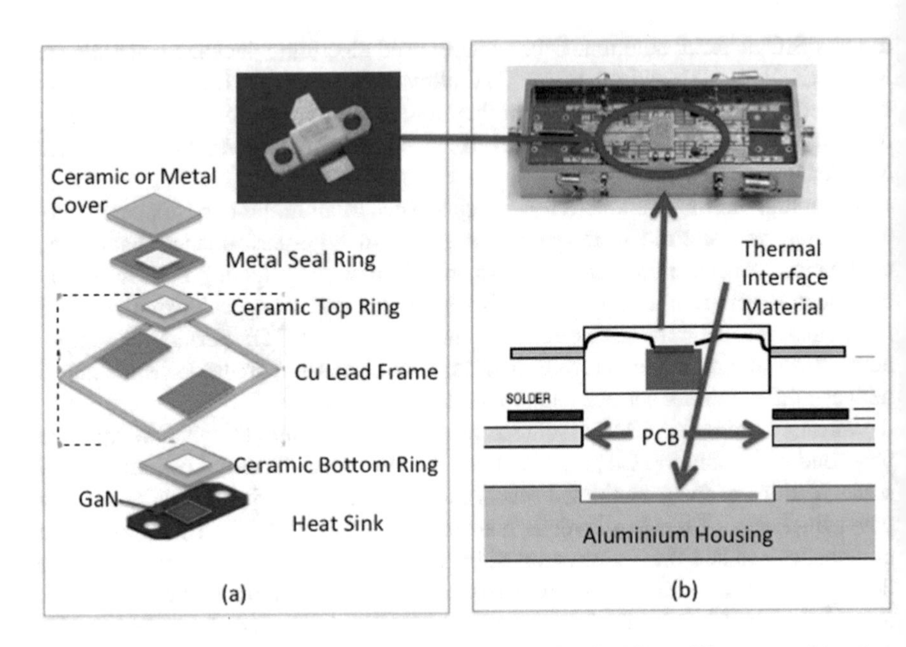

Fig. 9.1 (**a**) Schematic representation of RF power amplifier (**b**) RF amplifier mounted in a hermetic housing

(metallized alumina), copper or Alloy 48 lead frame, a metal seal ring, and ceramic or metal cover. Then the RF amplifier is placed in a housing. Since it is a lower cost material, light weight and machined very easily, aluminum is a widely used housing material. Unfortunately, the power amplifier needs to be mechanically attached to the housing due to CTE mismatch between the RF amplifier and the aluminum housing, instead of forming a joint with a high TC chemical bond interface. In addition, a thermal interface material, relatively low thermal conductivity material, needs to be used between the RF amplifier and the housing to maintain a continuous heat dissipation path. In order to eliminate the CTE mismatch problem, lower CTE materials such as Kovar and Titanium can also used be used and these materials can directly be bonded to the RF amplifier by soldering, silver sintering, or other methods. Unfortunately, Titanium and Kovar like other low CTE materials also have low thermal conductivity. These materials are also more expensive and heavier than aluminum. As a result, until recently when the application requires high heat dissipation aluminum still becomes the preferred choice. As we will discuss below, new developments in AlSiC and Aluminum Diamond manufacturing technology have started to provide excellent solutions to all the issues related to heat dissipation and CTE mismatch.

Both Aluminum Diamond and AlSiC are direct replacement materials for CuW, CuMo, and CMC due to their tailorable CTE as well as their high thermal conductivity. In fact, Aluminum Diamond provides at least 2× better thermal conductivity than the other Group 1 materials and it is an excellent choice for high performance demanding applications. In the past, the preference for copper-based composites as a heat spreader was due to the capability of creating the hierarchy in joining. Power amplifiers can be joined to RF packages using solders such as AuSn or AuGe because the melting point of the other joints was higher and they did not reflow at the processing temperatures of AuSn and AuGe soldering. The brazing material, copper/silver, used in copper composite-based RF amplifier manufacturing is not compatible with Aluminum Diamond and AlSiC due to its high processing temperature. As a result, in the past, copper composite-based heat spreaders were preferred to aluminum composite-based heat spreaders due to the lack of joining processes in the manufacturing of RF amplifier with aluminum composite-based heat spreaders. Development of nano-silver and silver sinter joining process has eliminated the advantage of copper composite-based heat spreaders. Power amplifiers can now be joined to aluminum based metal matrix composite using RF packages utilizing solders such as AuSn or AuGe. These process developments can now make Aluminum Diamond and AlSiC a drop-in replacement for the future RF amplifier. Therefore, performance demanding applications have started to replace copper composite-based heat spreaders with Aluminum Diamond without any modifications to their system design yet realizing the increased performance benefits. In fact, if required, it would also be possible to manufacture aluminum housings with a Aluminum Diamond or AlSiC region as an integral part of the aluminum housing. This would allow designers to place high thermal conductivity and controlled CTE material only in the high heat dissipation region. Such a design change would eliminate the necessity of manufacturing separate RF power amplifiers and possibly reduce the manufacturing cost.

Most diamond-based metal matrix composite materials used for thermal management applications are based on a primary metal (metals such as aluminum, copper, silver, or silicon) in combination with diamond particles. The size and volume fraction of the diamond particles are optimized to provide high thermal conductivity, low CTE, and good mechanical strength to MMC. Based on the MMC producer, these diamond particles can vary in size from 10 to 250 μm and are typically synthetic diamond with thermal conductivities between 1000 and 2000 W/m K. Testing has shown that the interface between the diamond particle and infiltrated metal is critical in the overall thermal conductivity and thermal stability of the MMC [2, 3]. The inability of the metal to fully wet the diamond particles can cause formation of micro voids or micro cracks at this interface which can lead to the degradation in thermal conductivity (as this interface is critical in the phonon transfer between the metal and diamond). During temperature cycling these micro voids (Figs. 9.2 and 9.3) can cause the interface to fail leading to a decrease in thermal conductivity.

For this reason, most diamond-based MMC producers utilize surface conversions or coatings on the diamond particles that promote proper wetting to minimize the formation of micro voids. For example, NMIC (an Aluminum Diamond manufacturer) utilizes a process that produces a thin reaction formed and diffusion bonded functionally graded SiC surface layer on the diamond particles [4]. Unlike a SiC coating, which would provide an additional thermal interface between the SiC layer and diamond particle [5] the SiC surface conversion layer is part of the diamond particle and thus eliminates the significant thermal resistance to the diamond. In

Fig. 9.2 Voiding in copper diamond

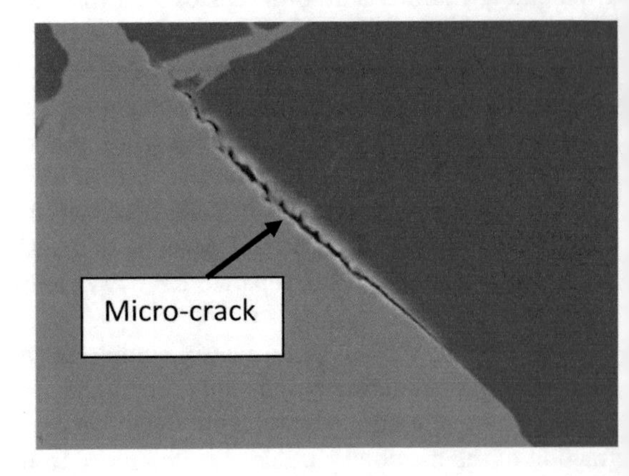

Fig. 9.3 Micro crack at copper to diamond interface

addition, the surface conversion on the diamond particle achieves minimal interface thermal resistance with the metal matrix which translates into good mechanical strength and stiffness, near theoretical thermal conductivity levels for the composite, complete wetting of the diamond particles by the aluminum due to the SiC surface layer, and the elimination of micro voids at the interface. Fig. 9.4 shows an SEM micrograph of non-converted and SiC surface converted diamond particles. Figure 9.5 shows an SEM image of Aluminum Diamond showing defect-free diamond particles to aluminum interface. As diamond composites are difficult to mechanically polish, argon milling was utilized to polish samples shown in Figs. 9.2, 9.3, and 9.5. The SEM images were then obtained using a JEOL field-emission scanning electron microscope [6].

To demonstrate the performance of the diamond to metal interface, temperature cycling tests are an important validation test to carry out [3]. For Aluminum Diamond, temperature cycling tests (−40 C × 30 min/120 C × 30 min, Air) were performed on samples (12.7 mm × 12.7 mm × 1.5 mm) that were fabricated using both SiC surface converted and non-converted diamond powder and produced via the squeeze casting process. The samples were exposed to 1000 temperature cycles

Fig. 9.4 SEM image of non-converted (*left*) and SiC converted (*right*) diamond particle

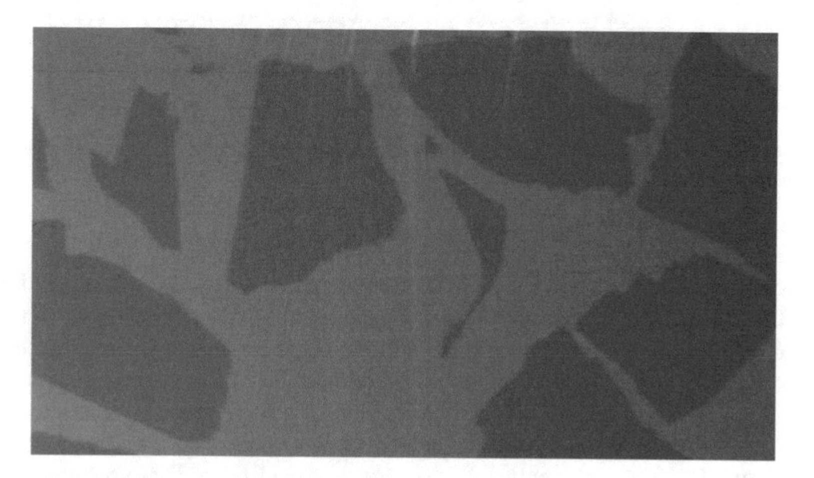

Fig. 9.5 SEM image of aluminum to diamond interface

and thermal conductivity was measured via laser flash method at the start of testing, 100 cycles, and 1000 cycles. Thermal conductivity of samples prepared with non-converted diamond powder quickly degraded while samples fabricated with SiC converted diamond powder maintained their thermal conductivity after each thermal cycle (Fig. 9.6). This test illustrates the importance of the metal to diamond interface and the role that SiC conversion plays in providing stable and consistent thermal performance for aluminum-based diamond composites.

Researchers and producers of copper diamond have utilized various coatings or addition of metal(s) to the copper matrix to promote proper wetting of the diamond particle and carbide formation at the diamond to copper interface. One paper outlines the use of Tungsten and Tungsten Carbide coatings [7] of the diamond particles (160–250 μm) to produce Copper Diamond composite by a Pulse Plasma Sintering infiltration process at 900 °C and 80 MPa applied pressure. Volume fraction of diamond was approximately 50%. During fracture tests for SEM imaging, the

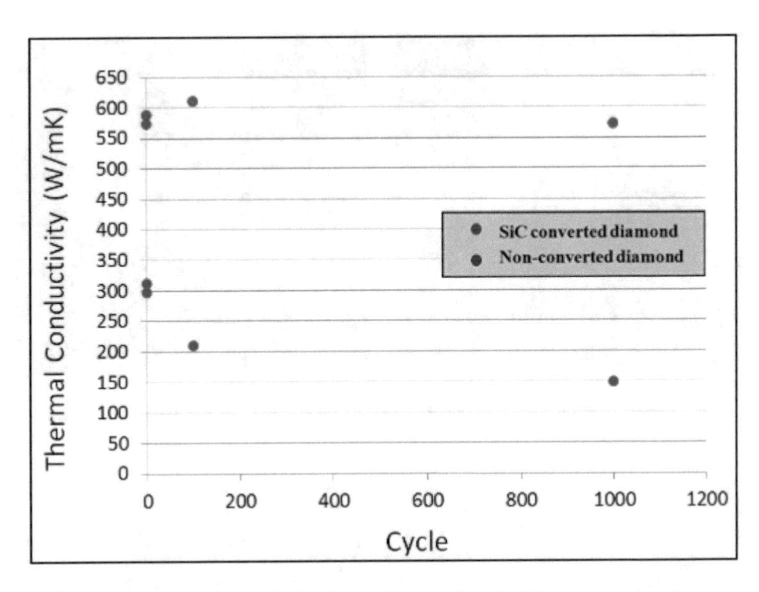

Fig. 9.6 Thermal cycling results of aluminum diamond with SiC converted and non-converted diamond powder. TC vs number of cycles is shown

authors noted that the Tungsten Carbide-coated diamond particles exhibited higher bond strength to the copper matrix than Tungsten-coated diamond particles as illustrated by the diamond fracturing on the Tungsten Carbide-coated diamond particles while the Tungsten-coated diamond particles fractured at the diamond to copper interface. This was reported to occur because the annealing process used to convert the Tungsten to Tungsten Carbide improved the adhesion of the coating to the diamond particles. Thermal conductivity reported for the Copper Diamond composite with Tungsten Carbide-coated diamond particles was 552 W/m K versus 686 W/m K for the Copper Diamond composite with Tungsten-coated diamond particles. The researchers did not report the coefficient of thermal expansion or the thermal cycling performance of the Copper Diamond composites produced, but it is the authors' opinion that the Copper Diamond composite produced with Tungsten Carbide-coated particles will have a more stable thermal conductivity after temperature cycling than the Tungsten-coated particles due to the stronger mechanical interface reported. Other researchers have reported fabrication of Copper Diamond composites utilizing the addition of alloying metals such as Titanium, Zirconium, Chromium, and other metals that are soluble in Copper to promote the wetting and bonding of Copper to the diamond particles. One study [8] reports the use of CuAgZr alloy mixed with 40–60 μm size diamond particles at a volume ratio of diamond of 40% with a reported thermal conductivity of 533 W/m K. The authors also used the Pulse Plasma Sintering Process (FAST) for infiltration, and densification was performed at 975 °C at 50 MPa. Coefficient of thermal expansion data was not reported, but the authors did indicate that thermal cycling was performed between −60 °C and 150 °C for 50 cycles with no measurable change in thermal conductivity.

The majority of Copper Diamond researchers and producers report utilizing copper powder as the copper source, hence a powder metallurgy process, but molten metal infiltration processes can also be utilized. The methods of densification employed vary from High Pressure and High Temperature Pressing, Hot Pressing, Hot Isostatic Pressing, and Pulse Plasma Sintering. The widespread adoption of Copper Diamond for RF applications has not occurred, most likely as a result of cost and limited manufacturers who are able to produce in volume.

Diamond-based composites, such as Aluminum Diamond, are beginning to play an important role in the advancement of microwave technology. Thermal limitations that have bottlenecked existing materials are being relieved by the use of high thermal conductivity Aluminum Diamond material (>500 W/m K). The higher thermal conductivity allows the user to see substantially lower junction temperatures (>25% reduction versus Copper Tungsten or Copper Moly Copper) or increased power outputs for their amplifier device. The typical use involves directly attaching the power amplifier (Gallium Nitride on Silicon Carbide or Gallium Nitride on Diamond) directly to an Aluminum Diamond heatsink/heat spreader using solders such as Gold Tin (AuSn), Gold Germanium (AuGe), or more recently nano-silvers. This allows the highest heat spreading from the power amplifier. The base of Aluminum Diamond heat spreader is then attached to a further thermal solution, such as a cold plate. In order to support the solder attach and avoid mechanically induced thermal stresses, it is important that the coefficient of thermal expansion (CTE) is a close match between the Aluminum Diamond heat spreader and power amplifier. Aluminum Diamond has a CTE of 7.0 ppm/K which is perfect match with GaN-based power amplifiers. The CTE of Aluminum Diamond is further controllable by the percentage of diamond present in the matrix. The low density of Aluminum Diamond (3.17–3.24 g/cm^3) is another important factor to users. Existing materials, such as Copper Tungsten (17.2 g/cm^3) and Copper Moly Copper (9.0 g/cm^3), have densities that are significantly higher than Aluminum Diamond and when being used in Aerospace or Satellite applications the weight differential when hundreds of devices are utilized becomes significant.

9.3 Electronic Package Fabrication Process

Reinforced metal matrix composites traditionally have been produced by various processes

1. Mixing metal and ceramic particles followed by sintering or hot pressing.
2. Dispersing the reinforcement particles in molten metal and then die casting, squeeze casting, or extruding or,
3. Infiltrating the molten metal into a preform (shaped reinforcement) with or without pressure.

Squeeze casting or gas pressure-assisted infiltration processes are most common methods to produce high TC Aluminum Diamond products. Since it is more mature product, all the processing techniques listed above can successfully be utilized for AlSiC products. However, in case of high TC Aluminum Diamond and AlSiC electronic package technology squeeze casting and gas pressure-assisted infiltration techniques are preferred to produce the products in the CTE range 4 ppm/K and 8 ppm/K range.

Not surprisingly, AlSiC electronic packages were first tested in the military electronics market largely due to high initial manufacturing costs. Following acceptance by the military market, demand has then increased for AlSiC metal matrix composites for use in commercial electronics. Following the same path, Aluminum Diamond electronic products have also been adopted first with the military electronics market, but the demand in commercial applications has started to increase rapidly, because the manufacturing process is the same for both products and this process was already accepted for AlSiC electronic products. The development of a flexible high volume manufacturing process for Aluminum Diamond is already in place. As more production volume is realized, manufacturing costs and lead time will continue to decrease. A flow chart of the current Aluminum Diamond and AlSiC component fabrication process is shown in Fig. 9.7.

The first fabrication step is the preparation of SiC and Diamond powders. This is followed by loading these powders either in a pre-shaped form or as loose powder into casting molds and infiltration of aluminum into the preforms. After the infiltration process, the Aluminum Diamond and/or AlSiC heat spreader or electronic housing package is cleaned, finished, and plated. For an electronic housing package, connectors are then soldered into the package to obtain the final product. The process is monitored and controlled in each processing step to assure product conformance to customer requirements. Product can be produced in a plate form and can be cut to customer required heat spreader/heat sink size or cast to net shape housing to meet customer design requirements. Composite heat spreaders and heat sinks are then solderable to an electronic housing prior to any connector attachment. A detailed production process including pressure infiltration casting process is described in articles published earlier [9–11], therefore will not be repeated here. Instead, we will focus on the differences between squeeze casting and gas pressure infiltration processes. For AlSiC or Aluminum Diamond, the squeeze casting process involves high pressure infiltration (50–150 MPa) of molten aluminum or aluminum alloy into tooling that contains an SiC preform, SiC powder, or SiC converted diamond powder depending on the material (AlSiC versus Aluminum Diamond)

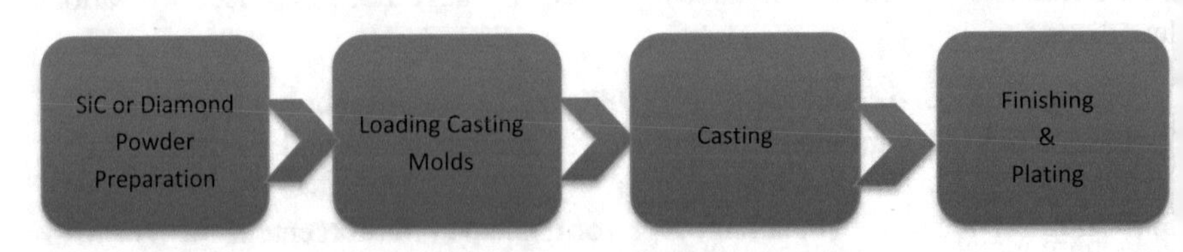

Fig. 9.7 Schematic AlSiC and aluminum diamond manufacturing process

and the process. The tooling is stacked to produce many plates which vary in size and thickness based on the customers final part dimensions. Parts are produced to net shape thickness. Squeeze casting achieves high pressure by mechanical compression of a molten metal into a sealed die containing the tooling. As it is mechanical in nature, squeeze cast pressure is higher than other processes, such as gas pressure-assisted infiltration. For the gas infiltration process, a similar configuration of tooling and powders are utilized, with the differences being that the process involves the melting of the aluminum or aluminum alloy during the casting run in a closed pressure vessel under vacuum and the infiltration of metal is performed by an inert gas at pressures between 7 and 10 MPa. The squeeze casting process is typically a faster process than gas pressure infiltration, as the melting of the metal is performed in separate equipment than that used for the infiltration process. For the gas pressure infiltration process, the melting of the metal is performed in the same equipment as used for infiltration. The melting of the metal can also be done separately and the molten metal can be carried into the infiltration vessel similar to squeeze casting. This difference makes the squeeze casting process a continuous operation while the gas pressure infiltration process is a batch operation. From a commercialization standpoint, both squeeze casting and gas pressure infiltration are proven high volume manufacturing processes and are both used extensively in the production of thermal management materials.

9.4 Core Capabilities

Core capabilities in relation to Aluminum Diamond/AlSiC manufacturing and characteristic features of the molten metal infiltration process are illustrated in Fig. 9.8, 3-D surface features are illustrated in Fig. 9.9, and a joining example is illustrated in Fig. 9.10. The net-shape casting ability is one of the fundamental advantages of the molten metal infiltration process. Net-shape parts are produced by infiltrating preforms which contain all the details of the finished part within a casting mold system. The casting molds are designed not only to contain the preforms, but to form the desired thickness of aluminum skin and all the dimensions of the final part. AlSiC plates 1.3–5.1 mm (0.050″–0.200″) thick and as large as 280 mm × 178 mm (11″ × 7″) are now in production. Aluminum Diamond parts can also be produced to the same size, but the current application does not require large size Aluminum Diamond plates. The standard Aluminum Diamond plate production is 0.51 mm–2.0 mm (0.020″–0.080″) thick and up to 51 mm × 102 mm (2″ × 4″) in length and width. Molten metal infiltration process is also capable of producing thinner [as thin as 0.25 mm (0.010″)] and thicker and larger baseplates. As-cast parts can easily meet ±0.127 mm (0.005″) in length and width and ±0.102 mm (0.004″) positional tolerance in any feature depending on the size of the part. Established production tolerance limits are given in Table 9.2.

A pure metal skin on the finished product is another characteristic of the molten metal infiltration process. An AlSiC and Aluminum Diamond microstructure with

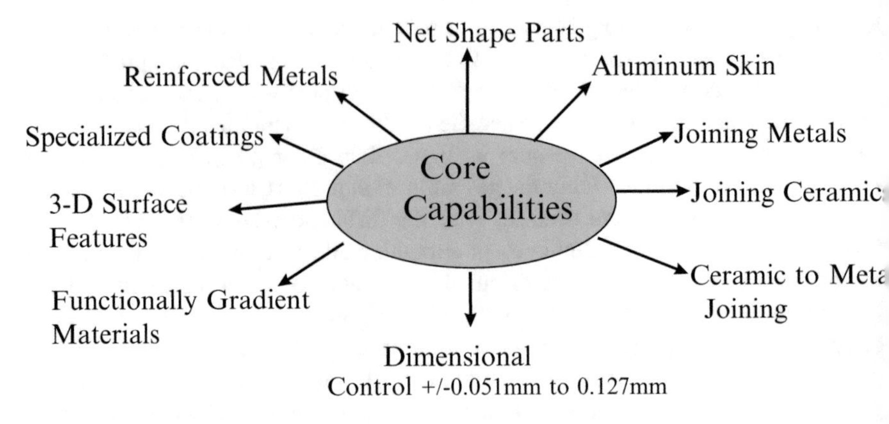

Fig. 9.8 Core capabilities in AlSiC and aluminum diamond production process

Fig. 9.9 3-D shaped aluminum diamond

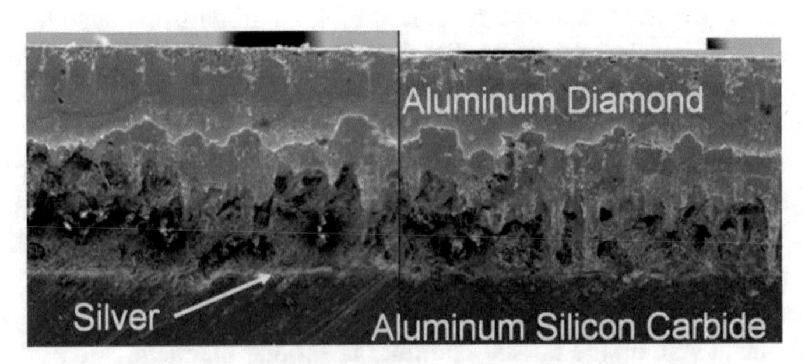

Fig. 9.10 Aluminum diamond joined to aluminum silicon carbide with nano-silver

Table 9.2 Dimensional design specifications

Item	Cut-parts	As-cast
Size tolerance (mm)	0.05	0.13
Thickness tolerance (mm)	0.05	0.05
Feature tolerance (mm)	0.05	0.10
Minimum hole size	0.30	0.30
Minimum distance from side surface to a hole	1.00	1.00
Max thickness (mm)	3.00	3.00
Tolerance of plating thickness (μm)	0.50	0.50
Surface roughness (μm)	<0.84	<0.84

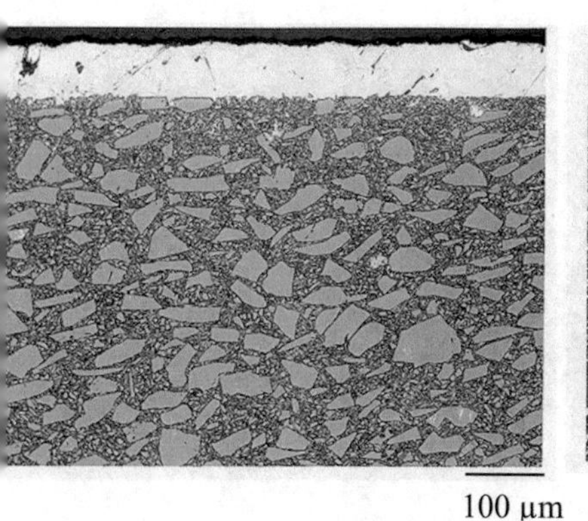

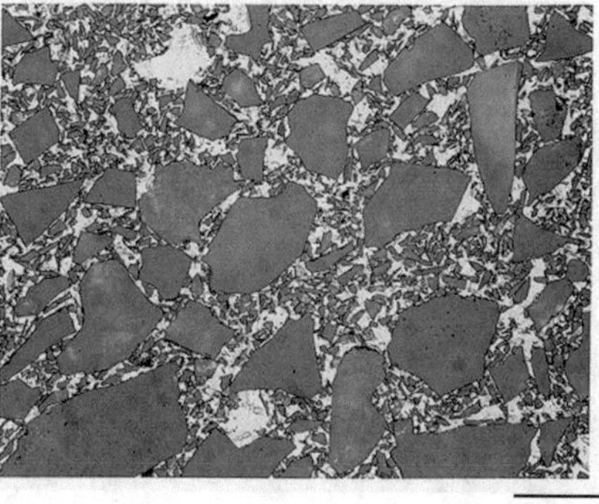

Fig. 9.11 Optical micrographs showing AlSiC microstructure with aluminum skin (*left*) and AlSiC microstructure with SiC particles dispersed in an aluminum matrix (*right*)

aluminum skin are illustrated in Figs. 9.11 and 9.12. The pure metal skin allows the part be plated or finished as though it were a pure metal. For example, there is no need for special Ni plating to activate and cover exposed SiC or diamond particles. Cast parts fabricated using molten aluminum infiltration process can be plated using traditional aluminum plating techniques. Application of various specialized coatings for specific solder alloys has also demonstrated. The optical micrograph in Fig. 9.13 illustrates such a coating designed for tin-lead solder. Since the coatings cost less than the Ni plating, these coatings easily replace Ni plating in applications where the parts will not be exposed to corrosive environments, or exposed only to mild corrosive environment that will not harm the parts. More recently, copper cold gas coating has been developed and also its use eliminated Ni plating requirements in soldering.

Fig. 9.12 SEM micrograph of aluminum diamond cross section showing the aluminum skin layer

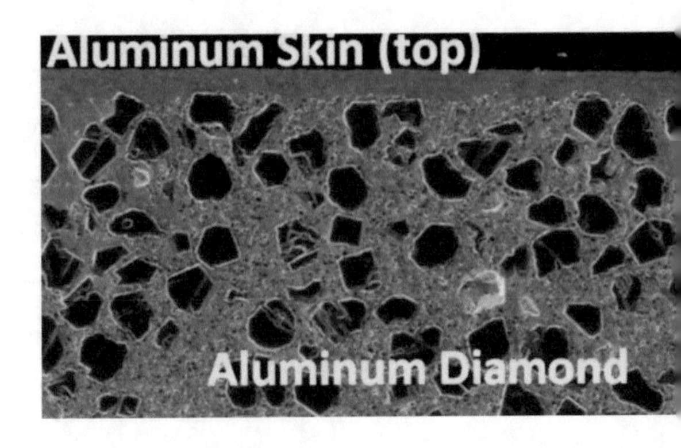

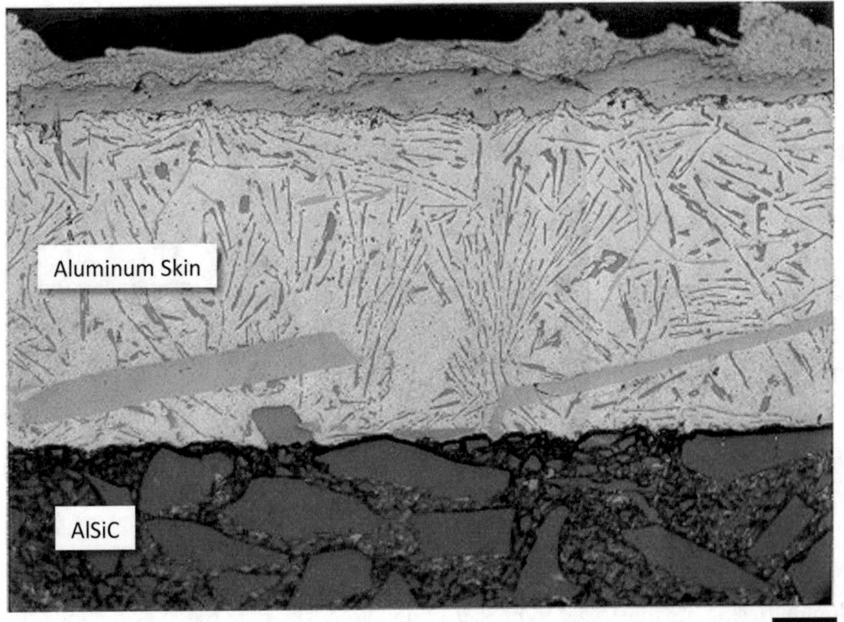

Fig. 9.13 An optical micrograph illustrating an AlSiC microstructure with aluminum skin and coating designed for tin-lead soldering. The second phase in the aluminum skin is Si particle dispersed in an aluminum alloy

9.5 Conclusion

It is predicated by these authors that Aluminum Diamond and Aluminum Silico Carbide will begin to displace traditional materials such as Copper Tungsten an Copper Moly Copper for RF microwave applications within the next 5 years as th heat dissipation from RF packages increases from the use of GaN, which is driving the need for higher thermal conductivity materials. The technical benefits and capabilities of these materials are mature to the point that they are a drop in replacement for existing materials. The amount of market displacement will be driven by the cost of these materials, which in volume is positioned to come down in price making the cost versus performance benefit trade-off a compelling choice for the end users.

References

1. Ray Pengelly, and RF Cree. 2009. GaN HEMT technical status: Transistors and MMICs for military and commercial systems. Research Triangle Park, NC 27709. IMS 2009 Boston, MA Presentation. Slide 10, 38 and 40.
2. Bollina, Ravi, and Sven Knippscheer. 2008. Advanced metal diamond composites—Love and heat relationship. *Electronics Cooling* 14(4): 20.
3. Loutfy, K., and H. Hirotsuru. 2011. Advanced diamond based metal matrix composites for thermal management of RF devices. *Wireless and Microwave Technology Conference (WAMICON), 2011 IEEE 12th Annual; 2011 Apr 18–19*, Clearwater Beach.
4. Pickard, S.M., et al. 2007. High thermal conductivity metal matrix composites. Patent 7,279,023 B2.
5. Johnson, W.B., and B. Sonuparlak. 1993. Diamond/Al metal matrix composites formed by the pressureless metal infiltration process. *Journal of Materials Research* 8: 1169–1173. doi:10.1557/JMR.1993.1169.
6. North Dakota State University, Electron Microscope Laboratory, https://www.ndsu.edu/ndsu-corelabs/electron-microscopy-center-people/. Accessed 5 Aug 2016.
7. Abyzov, Andrey M., Miroslaw J. Kruszewski, Lukasz Ciupinski, Marta Mazurkiewicz, Andrzej Michalski, and Krzysztof J. Kurzydlowski. 2015. Diamond-tungsten based coating-copper composites with high thermal conductivity produced by Pulse Plasma Sintering. *Materials and Design* 76: 97–109 .Journal homepage: www.elsevier.com/locate/matdes
8. Rape, A., X. Liu, A. Kulkarni, and J. Singh. 2012. *Alloy Development for Highly Conductive Thermal Management Materials Using Copper-Diamond Composites Fabricated by Field Assisted Sintering Technology*. Springer. Received: 18 April 2012/Accepted: 4 September 2012/Published online: 28 September 2012.
9. Sonuparlak, B. 1994. Aluminum MMC electronic materials produced by pressureless metal infiltration. In *Proceedings of Fifteenth and Sixteenth Annual Metal Matrix Composites Working Group Meetings*, 243–257.
10. Sonuparlak, B. and C. Meyer. 1996. Silicon carbide reinforced aluminum for performance packages. In *Proceedings of International Electronic Packaging Conference, IEPS Conference*, Austin, TX.
11. Sonuparlak, B., and D. Andrews. 1998. Silicon carbide reinforced aluminum for performance thermal management applications. In *PCIM International 98 Japan Proceedings*, Tokyo, 185–194.

Chapter 10
Advancement in High Thermal Conductive Graphite for Microelectronic Packaging

Wei Fan and Xiang Liu

10.1 Introduction

New electronic devices are constantly becoming more powerful and more compact. High power components, including radio-frequency (RF)/microwave (MW) electronics, diode lasers, light emitting diodes (LED), insulated gate bipolar transistors (IGBT), central processing units (CPU), etc., are utilized in a wide variety of industries such as telecommunications, automotive, aerospace, avionics, medical, illumination, and materials processing. These electronics generate great heat that must be dissipated, or else the electronics can be damaged by heat buildup. New capabilities are constrained by the ability of designers to remove heat in a cost-effective manner. Generally, every 10 °C increase in chip junction temperature decreases the lifespan of the device by half. The U.S. Air Force estimates that 55% of its electronics equipment failures are due to thermal effects [1].

Conventional thermal management products are typically constructed of either copper or aluminum. In many cases, heat generated by power electronics has exceeded the dissipation capability of aluminum, or even copper. The heavy weight of copper also makes it a less favorable choice for certain applications such as avionics and portable devices. In addition, copper and aluminum have very high coefficients of thermal expansion (CTE), compared with the semiconductor materials from which electronic components are constructed. A mismatch in CTE with electronic components causes thermal stress in the mounted electronic devices at an elevated temperature. This stress can cause unreliable operation and eventually lead to component failure. To combat this problem, heat sinks with direct contact with

W. Fan, Ph.D. (✉) • X. Liu, Ph.D.
Momentive Performance Materials, Inc.,
22557 West Lunn Road, Strongsville, OH 44149, USA
e-mail: Wei.Fan@momentive.com; weifan00@gmail.com

K. Kuang, R. Sturdivant (eds.), *RF and Microwave Microelectronics Packaging II*,
DOI 10.1007/978-3-319-51697-4_10

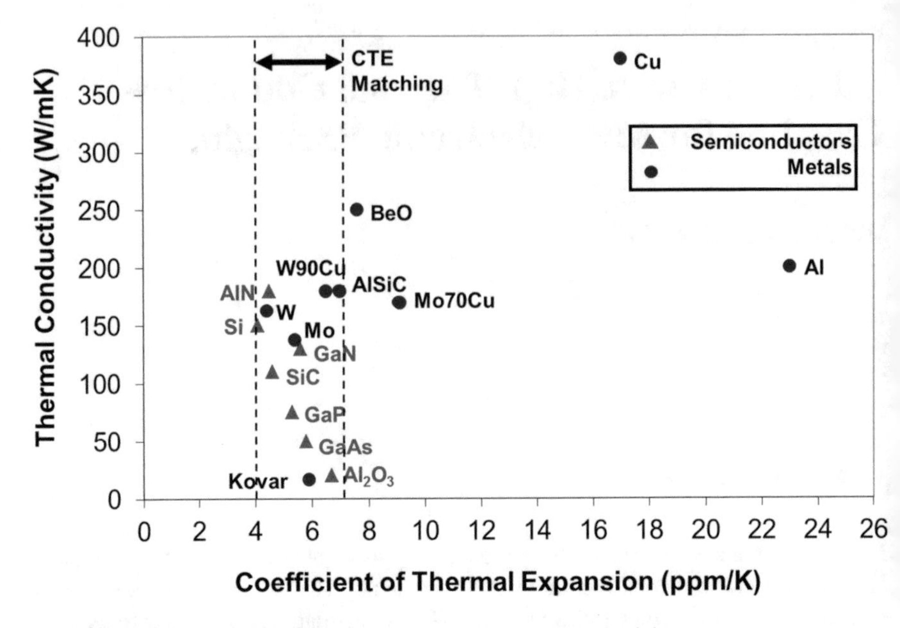

Fig. 10.1 Materials map for thermal expansion and thermal conductivity

semiconductor dies have been made from low-CTE materials such as aluminum silicon carbide (AlSiC), molybdenum-copper (MoCu), tungsten-copper (WCu), or copper-molybdenum laminates (CPC or CMC). However, these materials sacrifice thermal performance in exchange for better CTE matching with the electronic components. A comparison of thermal expansion and thermal conductivity of various materials is illustrated in Fig. 10.1.

In general, materials with high thermal conductivity, lightweight, and/or low thermal expansion are desired for high power thermal management application. In the rest of this chapter, we will discuss a group of advanced thermal management products based on high thermal conductivity and lightweight TPG graphite.

10.2 TPG and TPG-Metal Composites

10.2.1 Thermal Pyrolytic Graphite (TPG)

Thermal pyrolytic graphite (TPG*) is an advanced thermal management material discovered more than half century ago by Dr. Arthur Moore, a pioneer in graphite research at Momentive [2, 3]. In manufacturing process, the graphene planes are first laid on heated substrates in a high temperature chemical vapor deposition (CVD) process by pyrolysis of hydrocarbon gas. As illustrated in Fig. 10.2, the as-deposited pyrolytic graphite (PG) boards have a turbostratic, partially disordered structure and are further annealed into fully dense, highly ordered TPG at an

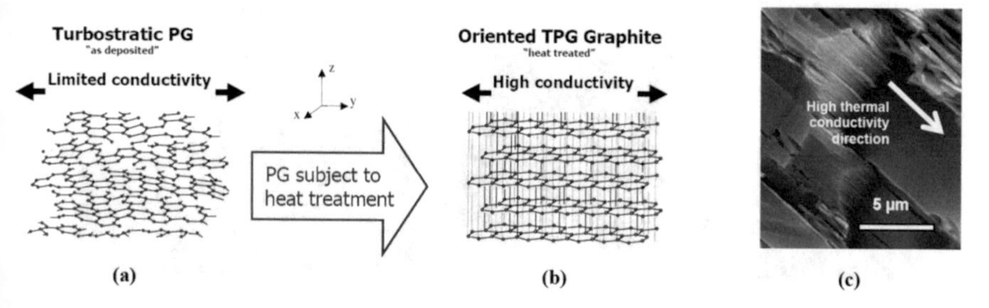

Fig. 10.2 Crystallinity illustrations of as-deposited PG showing turbostratic structure (**a**) and annealed TPG material with highly oriented graphene stacks (**b**), and electron microscope image of a cross section view of TPG material showing well-aligned graphene stack (**c**)

Table 10.1 Property comparison between commonly used thermal management materials

Material	Thermal conductivity (W/mK)	Coefficient of thermal expansion (10^{-6}/°C)	Modulus (GPa)	Specific gravity	Specific in-plane TC[a]
Aluminum	210	23.6	68	2.7	81
Copper	400	16.9	110	8.9	45
AlSiC	180	9.5	167	3.0	62
W85Cu	190	7.0	310	15.6	12
Mo70Cu	170	9.0	240	9.8	17
TPG Graphite	>1500 (in-plane)	0	30	2.25	650
	<10 (through-plane)	24	11		

[a]In-plane thermal conductivity divided by specific gravity
Typical properties are average data and are not to be used as or to develop specifications.

extremely high temperature. The well-aligned graphene planes created through this two-step process provide superior thermal conductivity (>1500 W/mK). Compared to copper, which is commonly used in passive cooling, TPG enables four times higher cooling power with one-fourth the weight (Table 10.1).

TPG is a relatively soft material, due to the weak Van der Waals force between the graphene layers. It can be machined into various shapes and sizes using regular machining tools and methods. A few examples of machined TPG tiles are shown in Fig. 10.3. Unlike most metals, which exert plastic deformation and produce chips or curl when being cut, TPG fractures into small graphite flakes which provide self-lubrication. Machining TPG usually consumes much less energy and, therefore, generates much less heat. Coolant is not necessary in TPG machining.

10.2.2 TPG-Metal Composite

The weak bonding between TPG layers tends to render conductive graphite particles when it touches other surfaces, which can raise contamination or electrical shorting issues. Also, the inertness of graphite prevents TPG from joining to other

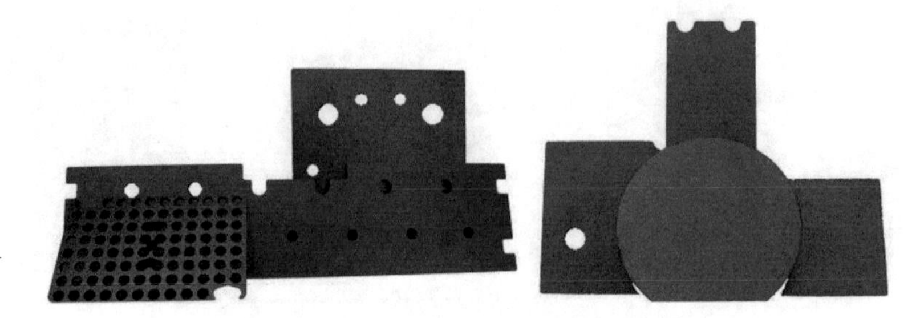

Fig. 10.3 Examples of machined TPG tiles with various shapes, sizes, and features

Fig. 10.4 TPG composites are made by encapsulating TPG in metal cladding. The composites can be machined for specific applications. Steps from left to right are (**a**) cutting TPG and inserting into metal clad; (**b**) bonding; (**c**) post-bonding machining

metals or ceramics through regular plating, soldering, and brazing processes. To eliminate the obstacle of direct use of TPG, a few novel bonding processes have been developed to encapsulate TPG into metal casing, such as aluminum and copper [4]. The TPG-metal composites are typically manufactured through a three-step process involving assembling, bonding, and machining as demonstrated in Fig. 10.4. The finished products behave like solid metal and can be further machined, plated, or bonded to other components to meet electronic packaging requirements.

Implementation of TPG composite products, such as TC1050* heat spreaders and TMP-FX thermal straps, into the heat path can quickly conduct the heat away from the sources and, therefore, greatly increase the electronics' efficiency and life. To address CTE-matching, TMP-EX heat sinks, which contain TPG material in various low-CTE alloys, were introduced to the market since 2010. Bonding TPG with CTE-matched alloys, such as WCu, MoCu, and AlSiC, simultaneously achieves high thermal conductivity from the TPG core and low thermal expansion from the metal encapsulation [5]. This TPG-metal composite heat sink provides high thermal conductivity, CTE matching, high reliability, and low weight density. This combination can significantly outperform the traditional monolithic incumbents using Al, Cu, WCu, MoCu, or AlSiC. TPG-metal composite products have been successfully deployed in the cooling systems of satellites, avionics, phased array radars, telecommunication, etc., which can take full advantage of its high thermal performance, high durability, and lightweight.

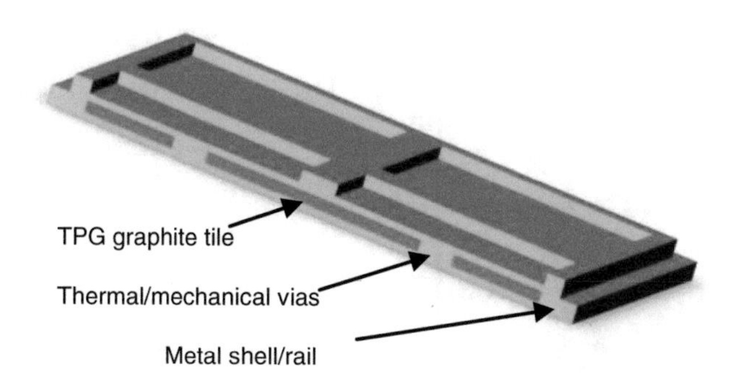

Fig. 10.5 Exposed view of a typical TPG-metal composite

10.3 Design, Property, and Reliability

Heat sink, heat spreader, or thermal strap removes high heat flux from critical electronic components. High thermal and mechanical performance as well as high reliability are basic requirements for any successful thermal management products. The TPG-based thermal management product portfolio, which includes TC1050 heat spreader, TMP-EX heat sink, and TMP-FX thermal strap, not only offers a wide array of advanced functionalities, but also meets the strictest electronic packaging standards.

10.3.1 Design Principles

As opposed to traditional graphite-metal or diamond-metal composites, TPG as a solid core material is fully or partially encapsulated by metal layers. Figure 10.5 illustrates the architecture of a typical TPG-metal composite. TPG provides the high thermal conduction and metal encapsulant renders the mechanical strength, hermeticity, machinability, platability, and solderability. This structure can eliminate the difficulties of machining, coating, and bonding often encountered when using conventional graphite- and diamond-metal composites. The TPG-metal composite, with higher thermal conductivity and lighter weight, can be processed and treated as its monolithic metal incumbents in the downstream assembling.

Mechanical features such as rails, bosses, vias, and pockets can be added to the TPG-metal composites when needed. TPG tiles as the core can also be created with various shapes and dimensions to maximize heat conduction and weight saving. One of the key factors to be considered in designing the TPG tiles is its anisotropic thermal property determined by the layered graphite structure. Unlike most of the thermal management metals or ceramics, TPG exhibits excellent in-plane thermal conductivity, but relatively poor through-plane value. Depending on the electronic packaging assembly design, TPG tile can be oriented to the direction(s) which aligns with the desired heat path.

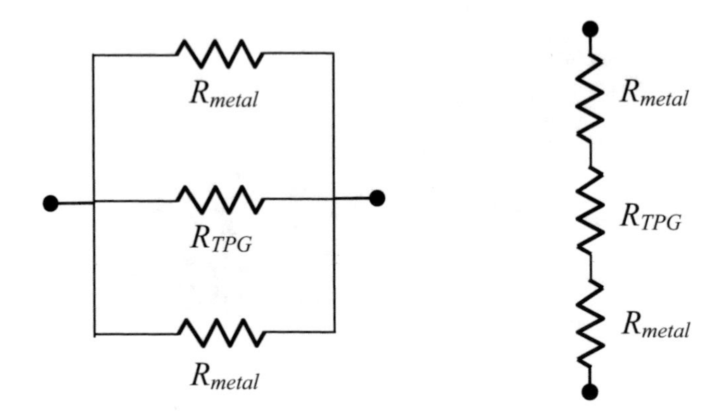

Fig. 10.6 Illustrations of in-plane heat conduction of three resistors connected in parallel (*left*) and through-plane heat conduction of resistors in series (*right*)

For simplicity, we can always view the TPG-metal composite as a 2D sandwich structure, which TPG core resides between two metal layers. The in-plane heat conduction can be considered as a circuit of three thermal resistors with metals and TPG connected in parallel, while the through-plane scenario consists of three resistors in a series path (Fig. 10.6). Therefore, the in-plane and through-plane thermal conductivities of TPG-metal composite can be estimated via the following two equations:

$$k_{\text{in plane}} = \frac{t_{TPG}k_{TPG} + t_{\text{metal}}k_{\text{metal}}}{t_{\text{total}}} \tag{10.1}$$

$$k_{\text{through plane}} = \frac{t_{\text{total}}}{\frac{t_{TPG}}{k_{TPG}} + \frac{t_{\text{metal}}}{k_{\text{metal}}}} \tag{10.2}$$

where k is thermal conductivity and t is thickness. The calculation is based on the assumption that interface resistance between TPG core and metal encapsulant is negligible, which is of great importance to receive expected high thermal conductivity, especially high through-plane thermal conductivity. This bonding interface issue is discussed further in the following session.

It is apparent that the in-plane and through-plane thermal conductivities of the TPG-metal composites are both functions of the ratio between TPG core thickness and metal encapsulant thickness. However, the in-plane and through-plane thermal conductivity functions follow distinct trends as shown in Fig. 10.7. To achieve significant thermal conductivity improvement in any TPG-metal composite designs, TPG core with a thickness 60% or more over the entire real estate is generally suggested.

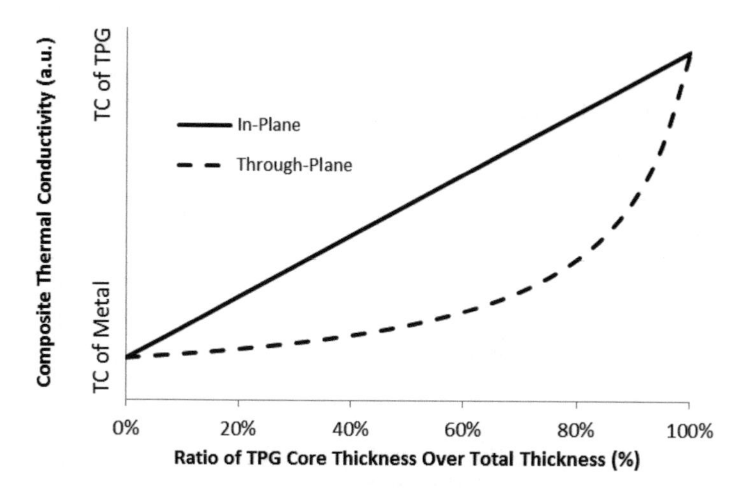

Fig. 10.7 Theoretical thermal conductivity of TPG-metal composite, showing the in-plane and through-plane values, follows two distinct functions

Fig. 10.8 Sketches of two common TPG tile orientations with graphene basal planes parallel to metal layers (*Configuration A*) and perpendicular to the metal layers (*Configuration B*), respectively

It is worth of reiterating that the thermal conductivity of TPG has two drastically different values, 1500 W/mK along the graphene plane and 10 W/mK perpendicular to the graphene plane. The alignment of the TPG high thermal conduction orientation with the desired heat flow direction is essential to the design. Two common TPG tile orientations are depicted in Fig. 10.8. Configuration A is suitable for long distance in-plane heat conduction or spreading situation, in which hot source is at one end and cooling mechanism is at the other end. When hot source and cooling mechanism are mounted on the opposite surfaces of a heat sink, Configuration B can dissipate heat through the thickness very well as well as along one in-plane direction.

The above principles are useful guidance to start any design work of TPG-metal composite or simply to select the types of TPG-metal products. The integration of TPG into a sophisticated thermal management system can be greatly benefited from computer thermal simulation using commercial software products, such as ANSYS, FloTHERM, COMSOL, SOLIDWORKS, etc. The product family of TPG-metal composite has been expanding continuously in the past 20 years to solve increasing thermal management challenges. Session 10.4 discusses the features and applications of each family group in detail.

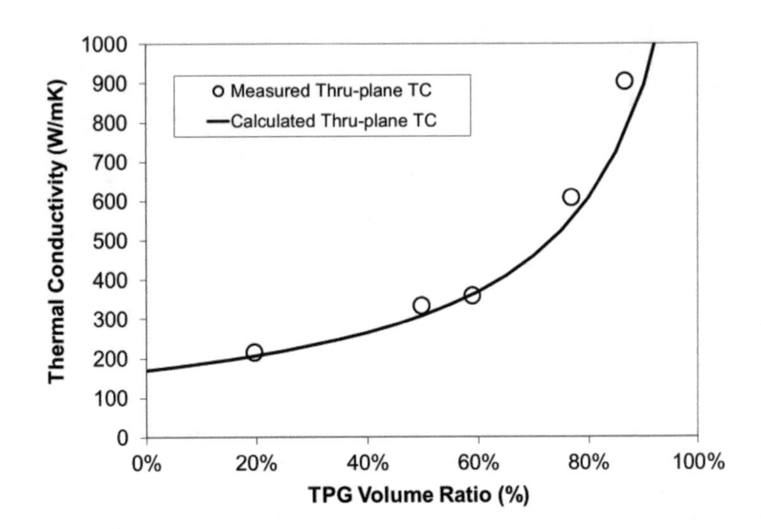

Fig. 10.9 The thermal conductivity of TPG-Mo70Cu composites at different graphite loadings. Note: test data. Actual results may vary

10.3.2 Thermal Conductivity

High thermal conductivity and low density are two important properties of all TPG-metal composites, including TC1050 heat spreader, TMP-EX heat sink, and TMP-FX thermal strap. A near-zero thermal interface resistance at the metal-TPG bond line is the key to realize the overall high thermal conductivity.

In one study, TPG-metal composites with Mo70Cu/TPG/Mo70Cu sandwich structure were prepared using Momentive's proprietary bonding technique. The TPG plates were cut in a way that all the graphene planes were aligned vertically to the metal substrates (Configuration B in Fig. 10.8), so that the heat flowing through the sandwich structure was conducted by the high thermal conductive TPG basal plane. The sample TPG loading varied from 20 to 87% in thickness (volume). The composite through-plane thermal conductivities were measured by using Netzsch NanoFlash LFA 447. As shown in Fig. 10.9, the measured through-plane thermal conductivity with various TPG loading matched the theoretical values, indicating an exceptionally low thermal interface resistance was achieved with this bonding method. The theoretical through-plane thermal conductivity was calculated using Eq. 10.2 based on the assumption that no thermal interface resistance presented. A thermal conductivity as high as 900 W/mK was demonstrated at the TPG volume loading of 87%.

10.3.3 Thermal Expansion

CTE value of the TPG-metal composition is mainly determined by the rigid metal encapsulants, especially when low-CTE metals, such WCu and MoCu which have exceptionally high modulus of elasticity, are chosen. This mechanism is the

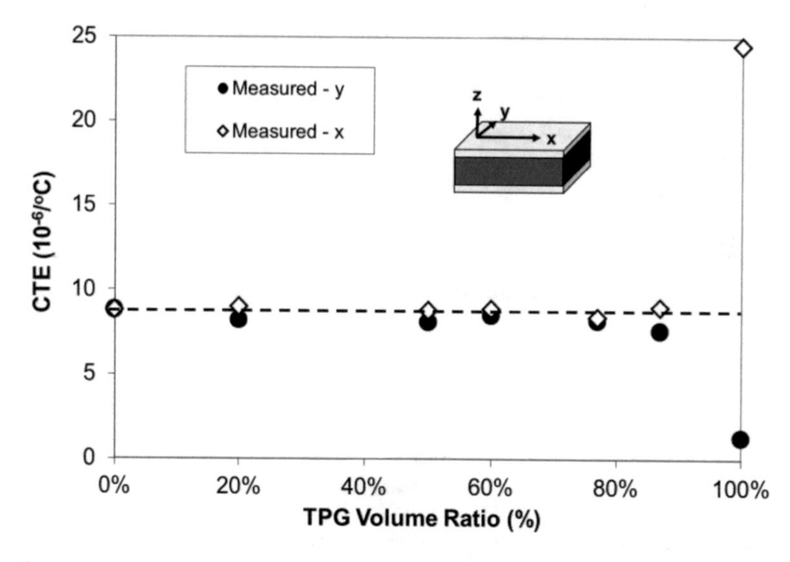

Fig. 10.10 The CTE of TPG-Mo70Cu composites in two in-plane directions as a function of TPG loadings. Note: test data. Actual results may vary

foundation for TMP-EX heat sink products to achieve the CTE matching property. A comparison of the modulus and CTE values among typical heat sink metals and TPG has been shown in Table 10.1.

CTE values of the TPG-Mo70Cu sandwiches, which were produced in the above thermal conductivity study, were characterized by using dilatometer Anter Unitherm 1251. The high thermal conductivity directions of TPG were oriented along y and z axis as labeled in Fig. 10.10. The measurement confirmed that the TPG-metal composites maintained the CTE of the metal shell in both lateral directions (x and y axis) and its thermal expansion is independent of TPG loadings (Fig. 10.10) and temperature (Fig. 10.11). Compared to the TPG material which has extremely low modulus, the rigid Mo70Cu encapsulant dictates the mechanical behavior of the composite and, thus, achieves low thermal expansion.

10.3.4 Reliability

In addition to the excellent TPG-metal bonding, shear strength of the metal–metal joint in the TPG-metal composite is on par or exceeds the metal ultimate yield strength. The exceptionally strong mechanical bond can help ensure the product reliability for years of service in harsh environments. Figure 10.12 shows the thermal conductivities of 20 pieces of TMP-EX heat sinks before and after a series of thermal cycling, vibration, and shock tests, following military microelectronics packaging standard MIL-STD-883H. TMP-EX heat sink is a TMP-metal composite product for chip-level thermal management and the discussion of its performance and application extends to the next session. As demonstrated in Fig. 10.12, no

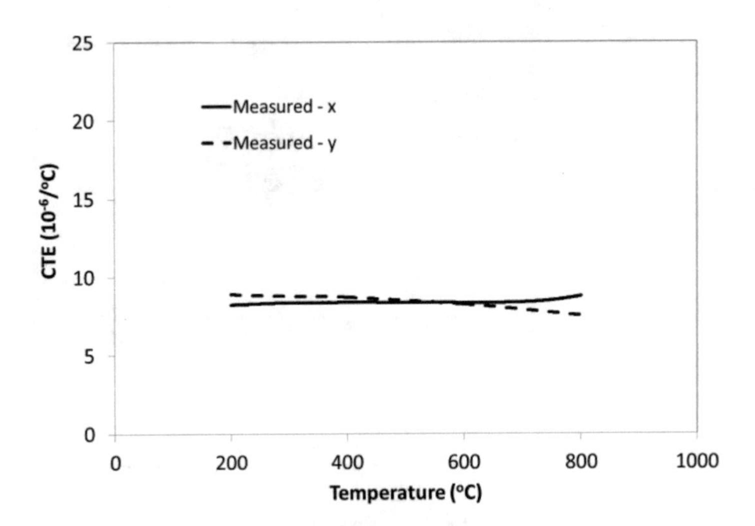

Fig. 10.11 The CTE of TPG-Mo70Cu composites in two in-plane directions as a function of temperature. Note: test data. Actual results may vary

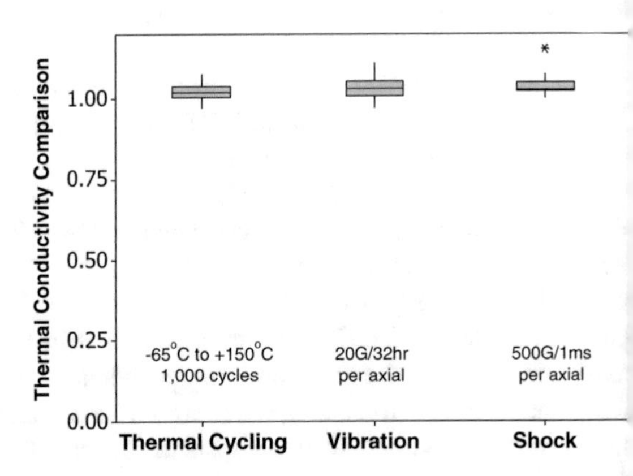

Fig. 10.12 Ratios of through-plane thermal conductivities of TMP-EX heat sinks measured before and after thermal cycling, vibration, and shock tests. Value 1.00 means no change. Note: test data. Actual results may vary

degradation in thermal characteristics of TMP-EX heat sinks was observed after each test. Furthermore, the TMP-EX heat sink regularly passed the strict hermeticity standard (helium leak rate < 10^{-8} atm·cm^3/s), guaranteeing no outgassing or graphite exposure in downstream assembling processes.

10.4 Applications

All three categories of TPG-metal composites, i.e., TC1050 heat spreader, TMP-EX heat sink, and TMP-FX thermal strap, inherit high thermal conductivity and lightweight from the TPG core. Individually, they were developed carrying different forms and properties to address different thermal management challenges and applications. The performance beacons of each product group are highlighted in Fig. 10.13.

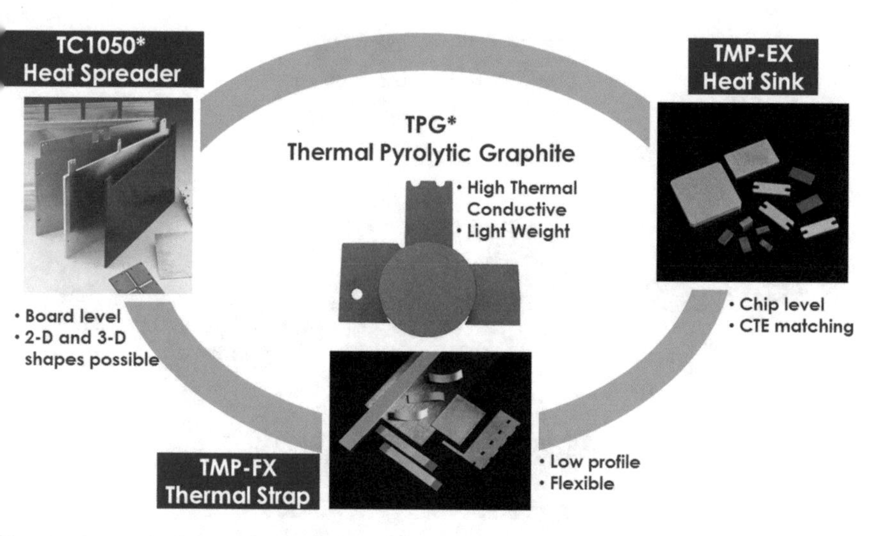

ig. 10.13 TC1050 heat spreader, TMP-EX heat sink, and TMP-FX thermal strap inherit high hermal conductivity and lightweight from the TPG core, but individually present different functionalities for specific applications

Take an example of a simplified electronic system as sketched in Fig. 10.14. Waste heat generated at high power electronic chip or die goes through a heat sink or flange and spreads out via a heat spreader backing the printed circuit board. The heat then dissipates to environment through convection cooling at the chassis. Alternative route may involve lid mount heat sink or thermal strap. The highlighted heat spreader, heat sink, and thermal strap can all be optimized with TPG-metal composites. Performance and application of TC1050 heat spreader, TMP-EX heat sink, and TMP-FX thermal strap are discussed separately in the follow sections.

10.4.1 TC1050* Heat Spreader

TC1050 heat spreader with its large format and high strength is commonly used for board-level thermal management and can bear some weight load. Typical applications include thermal cores for printed wire boards (PWB), avionic and satellite traveling wave tube (TWT) mounts, electronic chassis, and cold plates in radar systems, as well as thermal leveler for semiconductor wafer processing heaters or coolers. TC1050 heat spreader presents very high in-plane thermal conductivity, usually above 1000 W/mK and quickly carries the waste heat away from the high power devices. Typical size of TC1050 heat spreader is in the range of 20 cm × 10 cm and can go up to 60 cm × 60 cm. Instead of a simple 2D plate, TC1050 heat spreader very often contains features, such as fastener holes, risers, windows, trays, and rails, shown as examples in Fig. 10.15. Metal vias or buttons are added to strategic locations, as revealed in Fig. 10.15b, to increase the spreader strength as well as the through-plane thermal conductivity.

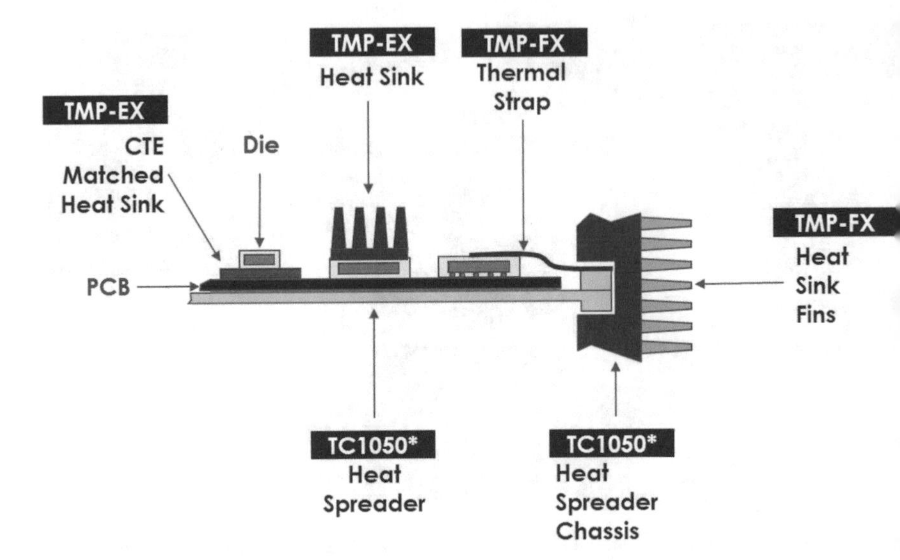

Fig. 10.14 Illustration of a typical electronic system, which presents a variety of opportunities for improving heat dissipation using TPG-metal composites

Fig. 10.15 TC1050 heat spreader examples (**a**) and cross section views (**b**)

In the demonstration shown in Fig. 10.16, heat load of 45 watt was added to the left side of an Aluminum 6061 heat spreader with a dimension of 300 mm × 100 mm × 3 mm and a TC1050 counterpart. The right side of each heat spreader was chilled with cooling liquid at 20 °C. In this comparison test, TC1050 heat spreader delivered 64% temperature reduction over the aluminum incumbent due to its exceptionally high thermal conductivity.

10.4.2 TMP-EX Heat Sink

TMP-EX heat sink is mainly developed for advanced thermal management of high power RF/MW microelectronics, laser diodes, and LEDs at chip level. Common die attach layouts can be categorized into two fundamental architectures, i.e.

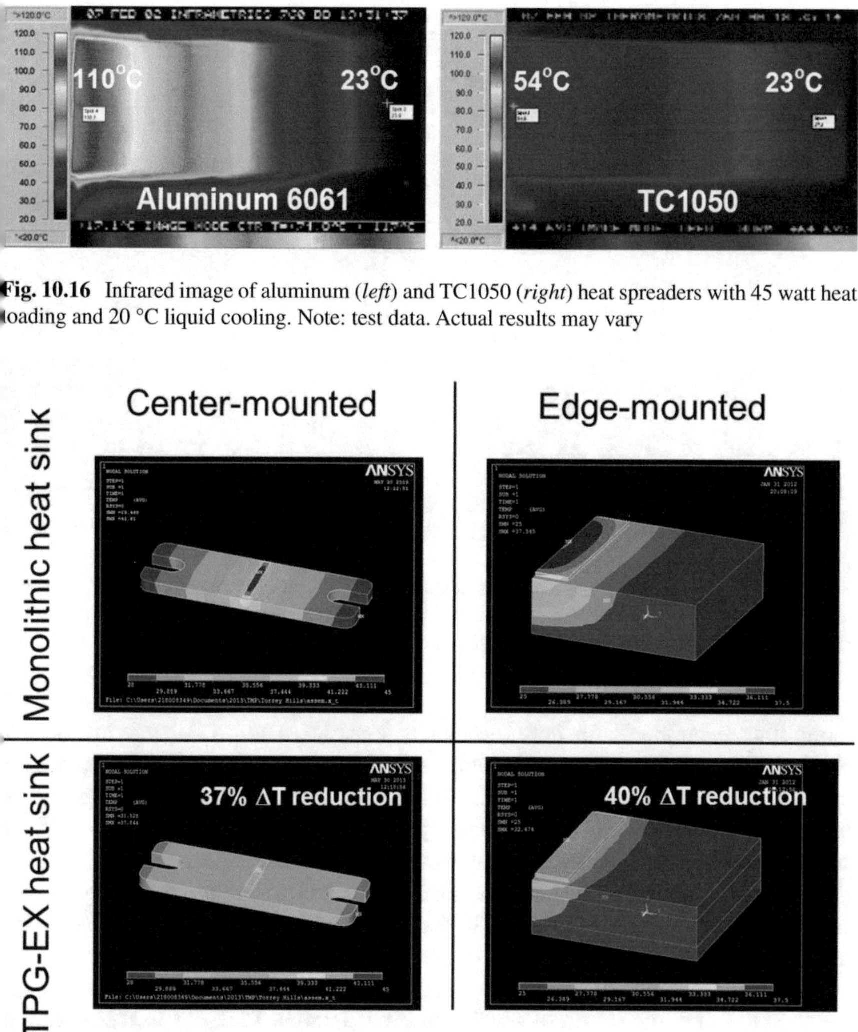

Fig. 10.16 Infrared image of aluminum (*left*) and TC1050 (*right*) heat spreaders with 45 watt heat loading and 20 °C liquid cooling. Note: test data. Actual results may vary

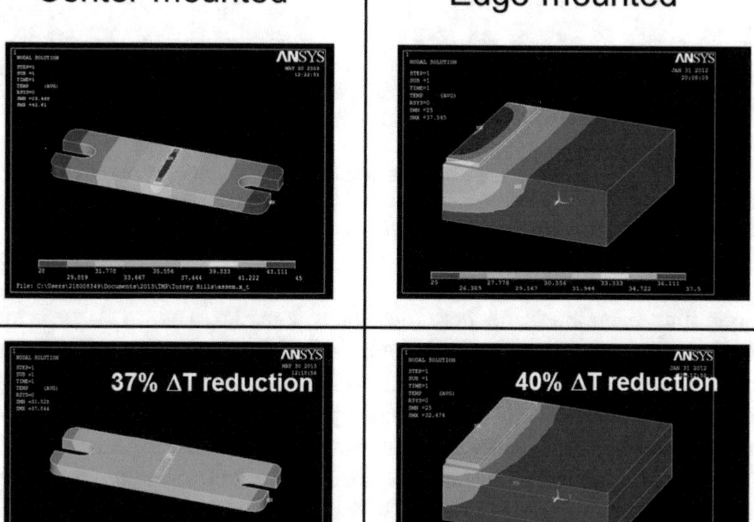

Fig. 10.17 Temperature profile comparison between monolithic W85Cu heat sink and TMP-EX heat sink with W85Cu encapsulation via thermal simulation

center-mounted and edge-mounted, as illustrated in Fig. 10.17. Most high power RF and MW devices are soldered onto the center of the CTE-matched heat sinks. In some optoelectronics applications, laser diodes are mounted on the edge of the heat sink to maximize light extraction. Cooling mechanism for both mounting architectures is always on the opposite surface of the die, which requires efficient in-plane heat spreading as well as high through-plane thermal conduction. Therefore, TMP-EX heat sink design very often chooses Configuration B in Fig. 10.8 to effectively conduct heat away from the source on top surface. Figure 10.17 presents a comparison of thermal performance between TMP-EX heat sinks and their

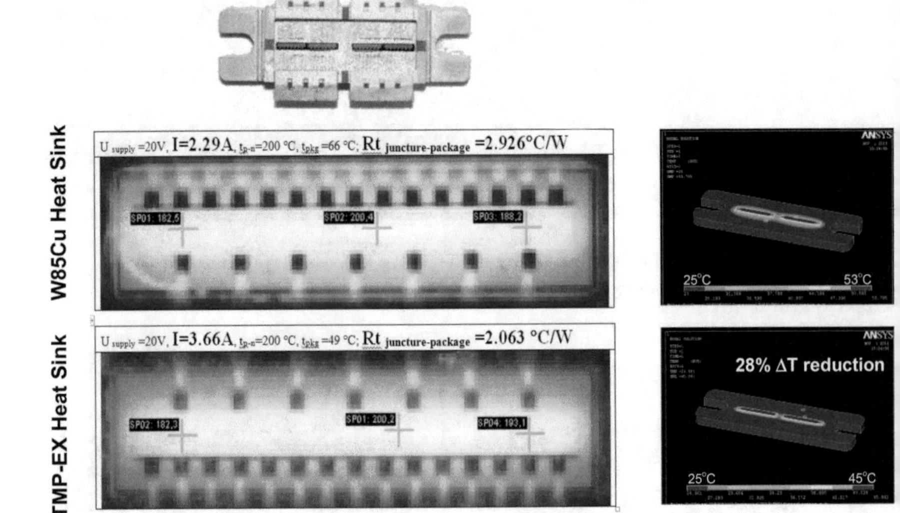

Fig. 10.18 *Top*—RF power transistors attached to TMP-EX heat sink; *Bottom left*—IR images of temperature profiles of power transistors using W85Cu heat sink and TMP-EX heat sink; *Bottom right*—simulated temperature profiles that match the measurement result (Courtesy of Syntec Microelectronics Ltd.)

monolithic incumbents for both die attach scenarios. Due to its superior thermal conductivity both in plane and through plane, TPG heat sink exhibits a much more uniform temperature profile, as well as a lower device temperature. The typical benefit of temperature reduction by using TPG heat sink is from 30% to 50%. In some extreme designs, 60% temperature reduction can be expected. The significant thermal resistance reduction and superior heat spreading power of the TMP-EX heat sink can be also translated to up to 80% more operating power from the electronic device without impacting operating temperature.

In an independent validation test for an RF application, LDMOS transistors were attached to a ceramic substrate which was brazed onto a TMP-EX heat sink and an incumbent W85Cu heat sink. The transistor temperature was determined via infrared imaging, shown in Fig. 10.18, and the TMP-EX heat sink exhibited 30% less thermal resistance than the monolithic W85Cu heat sink. As a result of the high heat conduction from TPG, the TMP-EX heat sink enabled 60% more power loading to the transistors without increasing the junction temperature. The heat reduction observed in this bench test also matched thermal simulation result very well.

10.4.3 TMP-FX Thermal Strap

TMP-FX thermal strap consists of TPG sheet sandwiched by very thin metal foils such as aluminum, copper, or tin. In addition to high thermal conductivity and low weight, the thermal strap also presents low profile, flexibility, formability, and solderability, which make it an excellent candidate for heat spreading in very tight space. Thanks to the flexibility of thin TPG sheet, TMP-FX thermal strap can be further shaped to create bends or turns without losing any of its high thermal conductivity. An example of shaped TMP-FX thermal strap is shown in Fig. 10.19.

Common applications of TMP-FX thermal strap extend to heat spreaders and heat sink fins. When used as heat sink fins, the high heat spreading power of TMP-FX thermal strap increases the effective heat dissipation area and thus improves the efficiency of finned heat sinks. In an automotive headlight application, 9 pieces of 10 cm long aluminum fins were used to remove waste heat from two high power LED chips operating up to 15 watts per chip [6]. TMP-FX thermal straps with identical dimension were later chosen to replace the aluminum fins. As discovered by thermal simulation (Fig. 10.20), TMP-FX fins presented less temperature gradient throughout the entire fin surface and removed more heat from the two devices. The thermal benefit was later validated by experiments. Measurement of the chip junction temperature (Fig. 10.21) confirmed that a 27% thermal resistance reduction of the entire system was achieved by replacing heat sink fins with TMP-FX thermal straps.

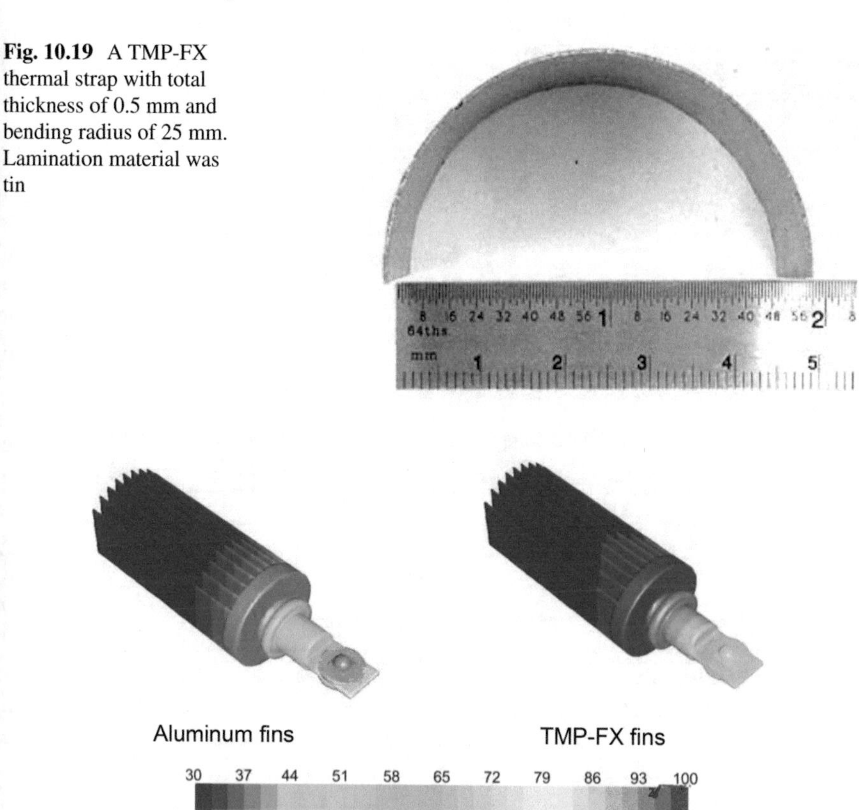

Fig. 10.19 A TMP-FX thermal strap with total thickness of 0.5 mm and bending radius of 25 mm. Lamination material was tin

Fig. 10.20 Simulated temperature profiles of aluminum finned heat sink (*left*) and TMP-FX finned heat sink (*right*) at 30 watts total power loading

10.5 Summary

The typical properties of TPG and its composites are compared to conventional thermal management materials in Table 10.2. Bonding TPG to metals simultaneously achieved high thermal conductivity and lightweight. An excellent thermal and mechanical bonding between the TPG and metal is the key to ensure desired thermal

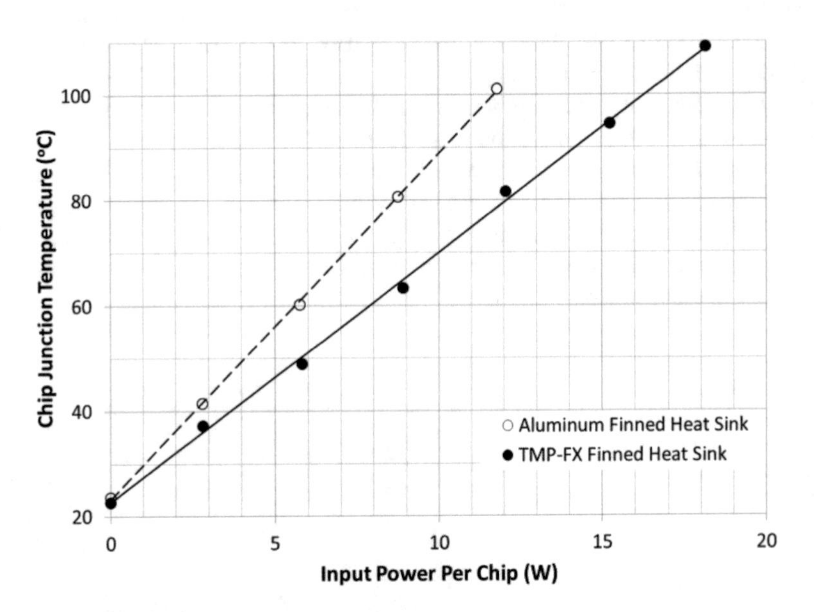

Fig. 10.21 Measured chip junction temperature as a function of input power per chip with the aluminum finned heat sink and the TMP-FX counterpart, showing a 27% temperature difference. Note: test data. Actual results may vary

Table 10.2 A comparison of TPG composites to the conventional thermal management materials

Material[a]	In-plane TC (W/mK)	Thru-plane TC[b] (W/mK)	Density (g/cm^3)
Aluminum	210	210	2.7
TPG + Aluminum	1073	507	2.4
Copper	400	400	8.9
TPG + Copper	1133	783	4.5
AlSiC12	180	180	3.0
TPG + AlSiC12	1060	435	2.5
W85Cu	190	190	15.6
TPG + W85Cu	1063	455	6.7
Mo70Cu	170	170	9.8
TPG + Mo70Cu	1057	416	4.8

Note: Typical properties are average data and are not to be used as or to develop specifications.
[a]Estimation is based on 67% of TPG loading
[b]TPG graphene plane is perpendicular to the metal surface for this estimation

and mechanical properties, as well as long-term reliability. Many high power electronics applications, which include RF/MW, laser, and LED, can benefit from the superior performances of TPG-metal composite products.

References

1. Pascoe, N. 2011. *Reliability Technology: Principles and Practice of Failure Prevention in Electronic systems*. Wiley.
2. Moore, A.W., A.R. Ubbelohde, and D.A. Young. 1964. Stress recrystallization of pyrolytic graphite. *Proceedings of the Royal Society of London. Series A: Mathematical and Physical Sciences* 280(1381): 153–169.
3. Ubbelohde, A.R., D.A. Young, and A.W. Moore. 1963. Annealing of pyrolytic graphite under pressure. *Nature* 198: 1192–1193.
4. Liu, X., W. Fan, and J. Mariner. 2010. TPG/TC1050 materials for the thermal management of electronics. *Advancing Microelectronics* 37(6): 20–26.
5. Fan, W., A. Rape, and X. Liu. 2014. How can millions of aligned graphene layers cool high power microelectronics. *International Symposium on Microelectronics* 2014(1): 433–437.
6. Fan, W., E. Galestien, C. Tomek, and S. Manjunath. 2016. Doubling the output of automotive LED headlight with efficient cooling using thermal pyrolytic graphite. In *15th IEEE ITherm*, 180.

Chapter 11
Carbon Nanotubes and Graphene for Microwave/RF Electronics Packaging

Xiaoxing Lu

11.1 Introduction

With the development of microelectronic processing technologies, electronic devices are constantly scaled down with better performance and lower cost, and as predicted by Moor's law, the density of transistors in integrated circuits would double every 18 months [1]. Leading semiconductor companies like TSMC and Intel are producing transistor with their 10 nm technologies. Further advance in scaling down is very challenging and almost approaching theoretical limit. For example, electrical resistivity of copper interconnects increases with the shrinkage of dimension due to grain-boundary and surface scattering [2]. Electromigration and hot spot of metal interconnects would also become big problems. On the other side, as the transistor is scaling down, more and more cooling is required. Effective thermal management and heat dissipation is becoming a more and more important topic in developing high performance and reliability semiconductor devices. Especially with the most recent development of three-dimensional (3D) integration, vertical dimension is expected to dramatically promote the integration density. However, long term reliability and thermal management issues will come up with higher integration density. These issues are especially important for microwave/RF microelectronics with high power consumption and heat generation. To address these issues, carbon nanotubes (CNTs) and graphene have been proposed as potential candidates for RF electronic packaging due to their ultra-high electrical conductivity, thermal conductivity and stability, resistance to electromigration, and mechanical strength [3–9]. CNTs and graphene have reputations of their excellent mechanical properties, thermal properties, and electrical properties. They are also highly stable and

X. Lu (✉)
Integra Technologies, Inc., El Segundo, California, USA
e-mail: luxx.ucla@gmail.com

K. Kuang, R. Sturdivant (eds.), *RF and Microwave Microelectronics Packaging II*,
DOI 10.1007/978-3-319-51697-4_11

resistive to heat and chemical attacks. In these regards, carbon nanotubes and graphene have been studied extensively from both academia and industry as RF packaging candidate materials. This chapter is going to present a review of carbon nanotubes and graphene advanced materials properties, potential applications, and challenges for RF packaging application.

11.2 Carbon Nanotubes and Graphene Structure and Properties

CNT and graphene as candidate RF packaging materials are highly investigated for their intrinsic properties, such as thermal properties, mechanical properties, electrical properties, rheological properties, microwave adsorption, environmental and toxicological impacts, effect of preparation, and gas barrier properties. These materials have been found owning high mechanical strength and modulus, high thermal conductivity with low contact resistance and small expansion coefficient, high electrical conductivity, high thermal and chemical stability and processing compatibility with thermal budget of front-end-of-line (FEOL) and back-end-of-line (BEOL). According to these studies, CNTs and graphene are very promising candidate materials for RF packaging applications.

11.2.1 Structure of Carbon Nanotubes and Graphene

The atomic number of carbon is 6, and atom electronic structure is $1s^2 2s^2 2p^2$ in atomic physics notation. When carbon atoms combine to form graphene, sp^2 hybridization occurs. In this process, one s-orbital and two p-orbital combine to form three hybrid sp^2-orbitals at 120° to each other within a plane. This strong covalent σ bond binds the atoms in the plane and results in high stiffness and high strength of a CNT. The remaining p-orbital is perpendicular to the plane of σ bond. It contributes mainly to the interlayer interaction and is called the π bond. The interlayer interaction of atom pairs on neighboring layers is much weaker than an σ bond.

Graphite is 3D layered hexagonal lattice of carbon atoms. A single layer of graphite forms a 2D material, called a graphene layer. The layer–layer separation of 3.354 Å is much larger than nearest-neighbor distance between two carbon atoms, $a_{c\text{-}c}$ = 1.42 Å. A single-wall carbon nanotube (SWCNT) is best described as a rolled-up tubular shell of graphene sheet which is made of benzene-type hexagonal rings of carbon atoms. The body of the tubular shell is thus mainly made of hexagonal rings (in a sheet) of carbon atoms, whereas the ends are capped by half-dome-shaped half-fullerene molecules. A multi-wall nanotube (MWCNT) is a rolled-up stack of graphene sheets into concentric SWCNTs, with the ends again either capped by half-fullerenes or kept open.

A nomenclature (*n*, *m*), used to identify each single-wall nanotube, refers to integer indices of two graphene unit lattice vectors corresponding to the chiral vector of a nanotube. Chiral vectors determine the directions along which the graphene sheets are rolled to form tubular shell structures and axis vectors perpendicular to the tube. The chiral vector *C* can be expressed as $C = na_1 + ma_2$, where a_1 and a_2 are the unit cell base vectors of the graphene sheet, and $n \geq m$ (Fig. 11.1). While $m = 0$, the structure is zigzag tubes, and $n = m$ for armchair. Since $a_1 = a_2 = 0.246$ nm, the nanotube circumference (*L*) (the magnitude of *C*), diameter, and the chiral angle (θ) in the nanometers are defined as [10]:

$$L = |C_h| = a\sqrt{n^2 + nm + m^2} \tag{11.1}$$

$$d_t = L/\pi \tag{11.2}$$

$$\theta = \sin^{-1} \frac{\sqrt{3m}}{2\sqrt{n^2 + nm + m^2}} \tag{11.3}$$

The nanotubes of type (*n*, *n*) are commonly called armchair nanotubes because of the _/‾_/ shape, perpendicular to the tube axis. Another type of nanotube (*n*, 0) is known as zigzag nanotube because of the \/\/ shape perpendicular to the axis. All the remaining nanotubes are known as chiral or helical nanotubes and have longer unit cell sizes along the tube axis.

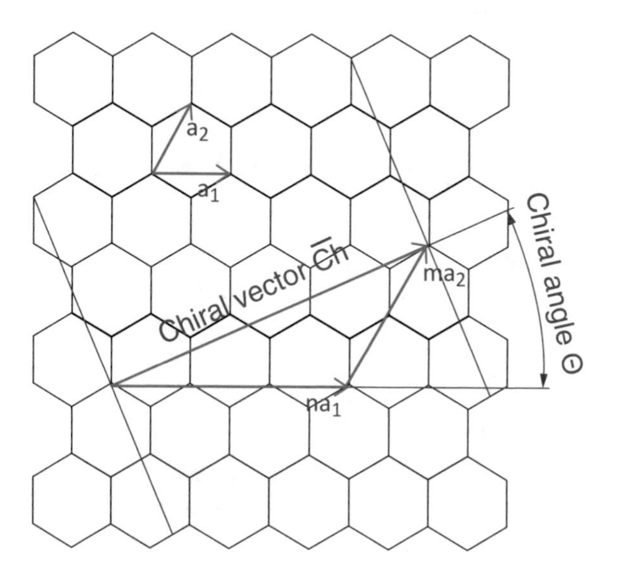

Fig. 11.1 Chiral angle and vector representation (Chiral vector $C = na_1 + ma_2$, where a_1 and a_2 are the unit cell base vectors of the graphene sheet.) [11]

11.2.2 Mechanical Properties

The discovery and potential wide range of applications of CNTs and graphene have stimulated many additional studies of the mechanical properties of these novel fibers. Theoretically, simple deformation experiments could determine the

Table 11.1 Data summary on CNTs Young's modulus-E (assuming an effective thickness of 0.335 nm for SWCNTs)

Authors	Method	*E* (TPa)	Refs.
Treacy et al.	TEM	1.77	[24]
Yakobson et al.	TEM and atomistic simulations	1.08	[12]
Wong et al.	AFM	1.28	[25]
Lu	Empirical force-constant model	0.97	[26]
Krishnan et al.	TEM	1.25	[27]
Hernandez et al.	Non-orthogonal TB	1.24	[20]
Gao et al.	MD	0.96	[22]
Salvetat et al.	AFM/TEM	0.81	[28]
Yu et al.	AFM/TEM	1.002	[29]
Tombler et al.	AFM	1.2	[30]
Demczyk et al.	TEM	0.9	[31]
Li and Chou	CM/FEM	1.05	[23]
Xiao et al.	CM	1.13	[32]
Kirtania and Chakraborty	CM/FEM	1.004	[33]
Ding et al.	TEM/AFM	0.94	[34]
Lu and Chen	CM/FEM	0.881	[35]
Avila and Lacerda	MM/FEM	0.994	[21]
Huang	AFM	0.97	[36]
Rossi and Meo	MM/FEM	0.916	[37]
Faccio et al.	QM	1.01	[18]
Lu and Hu	FEM	1.024	[11]

Table 11.2 Data summary on graphene Young's modulus-E (assuming an effective thickness of 0.335 nm for SWCNTs)

Authors	Method	*E* (TPa)	Refs.
Meo and Rossi	MM/FEM	0.958	[38]
Blakslee et al.	Resonance and static test	1.02	[39]
Kirtania and Chakraborty	CM/FEM	0.998	[33]
Liu et al.	Ab initio	1.05	[19]
Brunch et al.	AFM	1.0	[40]
Lee et al.	AFM	1.0	[17]
Faccio et al.	QM/MM/DFT	0.96	[18]
Shokrieh and Fafiee	CM/FEA	1.04	[41]
Lu and Hu	MM/FEM	1.132	[11]

corresponding mechanical properties: Young's modulus *E* can be measured from a tensile test that characterizes solid resistance to uniaxial stretching and is defined as the ratio of stress to strain. A great number of experimental and theoretical studies have been undertaken to estimate their mechanical properties. Experimental methods for measuring the mechanical properties of CNTs are mainly based on transmission electron microscopy (TEM) [12–16] and atomic force microscopy (AFM) [17]. In addition to experimental endeavors, theoretical and numerical evaluations of the material properties of CNTs have also been undertaken through density function theory (DFT) [18] including Ab initio [19] and tight-bonding (TB) [20], quantum mechanics (QM) [18], molecular mechanics (MM) [21], molecular dynamics (MD) [22], equivalent-continuum modeling (CM) [23] approach, and finite element method (FEM) [23].

The results of Young's modulus-E have a big variation and the differences probably arise through different experimental measurements and simulation assumptions. A rough estimate of CNTs and graphene's mechanical properties would be ~130 GPa for tensile strength and ~1 TPa for Young's modulus [11], which is 100 times stronger than the strongest steel. The results from literatures are listed in Tables 11.1 and 11.2. The high Young's modulus provide CNTs and graphene with high reliability under high mechanical and thermal stress during RF electronic device operation.

11.2.3 Electrical Properties

The electronic properties of the graphene strongly influence those of the carbon nanotubes. Two-dimensional graphene sheet is a zero-gap semiconductor. The dispersion at the Dirac points is circular, independent of direction. Electrons or holes can flow in all directions and there is no bandgap. When it is rolled up along armchair directions, a metallic CNT (m-CNT) is obtained, whereas other directions can give a semiconducting CNT (s-CNT) [42, 43]. Smaller CNTs have lower carrier density and larger carrier localization, thus larger CNTs usually have a larger electrical conductivity. CNTs have no Peierls instability—a metal-to-insulator transition, which is a common issue for most 1D materials when they are sized to nano-scale [44]. This advantage makes CNTs a potential material for ultra-fine interconnects. As long as CNT length is smaller than electron mean free path, electrons transport ballistically in CNTs. Due to extremely weak acoustic phonon scattering along the axis, ballistic transport can reach micrometers under low bias for a perfect metallic CNT.

CNT can carry a current density up to 10^9 Acm^{-2}, which corresponds to about 25 μA of current for a single metallic CNT [45, 46]. Open-end MWNT has been found to have large current-carrying capacity. The conductance will be greatly improved by integrating open-ended MWNTs with metallic electrode contact. Naeemi et al. claimed MWNTs interconnects conductivities are several times larger than Cu based on physical models [47]. In reality, there are always defects presenting in CNTs structures, and electrical conductance may be lower due to electron scattering [48–50], intershell/interwall coupling [51], tube–tube contact resistance [52], and contact resistance between CNTs and other materials [53–55].

Graphene has been long thought to be impossible until its first isolation in 2004 by Andre Geim and Konstantin Novoselov at the University of Manchester [56]. Due to its 2D nature and honeycomb atomic structure, electrons are able to move and behave as massless in these materials. The effective speed of electrons in graphene is about 300 times less than the speed of light in vacuum with a remarkable reported values in excess of 200,000 $cm^2 V^{-1} s^{-1}$ electron mobility at room temperature [57]. The ultra-higwh fast electrons provide a high conductivity of graphene. The corresponding resistivity of graphene is about 10^{-6} Ω cm and much less than the resistivity of silver, which is the lowest at room temperature. The electrons in graphene can travel large distances without being scattered, making it a promising material for very fast electronic components. The electrical properties of graphene have proven to be better than many related compounds, such as carbon nanotubes, because of its high surface area and electrical conductivity.

11.2.4 Thermal Properties

High thermal conductivity (κ) is one of the most attractive intrinsic properties of CNTs [58]. There are many theoretical and experimental studies on the thermal transport property of CNTs [59–86]. Berber et al. predicted thermal conductance κ of an (10, 10) SWNT to be ~ 6600 $Wm^{-1} K^{-1}$ at room temperature based on molecular dynamics simulations [69]. The high thermal conductivity (κ) of CNTs is attributed to both electrons and phonons due to their large mean free paths in CNTs at room temperature. Experiments measured κ values of individual CNTs vary from 42 to 3000 $Wm^{-1} K^{-1}$, depending on tube diameter [69, 76], length [65, 71], chirality [74], defect [78, 81], temperature [63, 66], etc. For MWNTs, heat transport is more complex due to strong intershell interaction. Kim et al. claimed a thermal conductivity of individual MWNTs was >3000 $Wm^{-1} K^{-1}$ at room temperature using suspended MWNTs (~1 μm). Choi et al. reported thermal conductivity to be 650–830 $Wm^{-1} K^{-1}$ for individual MWNTs by employing a 3ω method [71, 85]. Yi et al. presented thermal conductivity of millimeter-long aligned MWNTs to be only about 25 $Wm^{-1} K^{-1}$ due to large number of defects and increased to a high value after annealed at 3000 °C. For practical applications, collective thermal transport property of CNTs aggregates (carpet, mat, yarn…) is of more interest than the intrinsic κ of individual CNTs and collective thermal transport property of 2D and 3D random CNTs networks. The reported values of a vertical aligned CNTs (VACNTs) array vary from 1 to 250 $Wm^{-1} K^{-1}$. The big variation of reported values from literatures might origin from the intrinsic differences in CNTs including quality (defect density), structure (diameter, chirality and wall layer numbers), substructure (Swiss roll, Russian roll or Bamboo roll) and so on. Some other differences like CNTs alignment, tip entanglement, packing density, adhesion to synthesis substrate, catalyst residual, measurement techniques, sampling processes, and calculations model and assumptions may also generate different results.

Thermal transport in graphene has attracted attention due to the potential for thermal management applications. A high value of 5300 W $m^{-1} K^{-1}$ [87] has been reported at early studies. Later studies on graphene thermal conductivity show its values varying from 1500 to 2500 W m^{-1} K^{-1} [88–91]. In addition, thermal conductivity can be reduced to about 500 W m^{-1} K^{-1} at room temperature on amorphous materials and polymeric residue because of graphene lattice scattering by the substrate [92–94]. The variations in the reported thermal conductivity values can be attributed to large measurement uncertainties and tested graphene intrinsic differences. The thermal conductivities of most widely used electronic packaging materials are 400 W/mK for Cu, 430 W/mK for Ag, and 250 W/mK for Al. Graphene and CNTs have much higher thermal conductivities than those metals and even the best bulk crystalline thermal conductor, diamond. Therefore, they should be a very good substitute to traditional packaging materials.

11.3 Applications

11.3.1 Carbon Nanotube for Thermal Interface Materials (TIMs)

TIM is a layer of materials between two solid blocks to reduce the interfacial thermal resistance because of the surface waviness and roughness, and provide certain adhesion strength. As shown in Fig. 11.2, TM1 and TM2 are thermal interface material between backsides of the chip and the heat spreader/cap, the heat spreader and the heat sink, respectively. TIMs are important thermal solutions and have been extensively studied in microelectronic industry [95–111]. Conventional TIMs include polymer TIMs filled with high thermal conductive fillers, folder TIMs, and phase change materials (PCM). With transistor scaling-down and increasing intensity, higher cooling capacity is in demand. However, the conventional TIMs and related technologies have more or less reached their limits due to their low thermal conductivity, increasing bond line thickness with high thermal conductive filler loader, high coefficient of thermal expansion (CTE), low compliance, high cost, low fatigue strength, and low reliability. It is highly demanded to develop high-performance TIMs for next-generation RF electronic packaging.

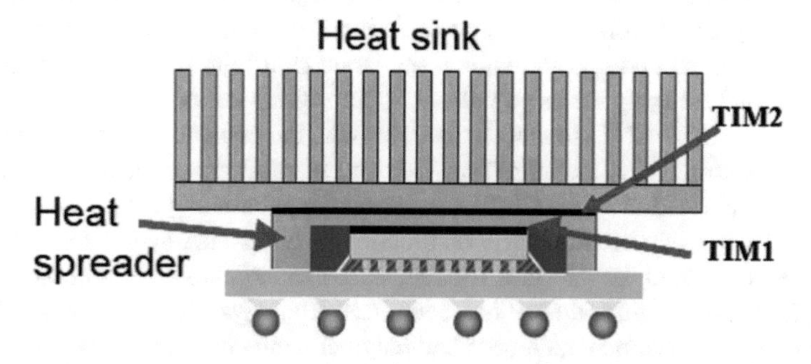

Fig. 11.2 Schematic illustration of TIM architecture in flip-chip technology [112]

The development of CNT TIMs has attracted great interests due to CNTs' high thermal conductivity, low CTE, high thermal and chemical stability and flexibility. However, it is found CNT/polymer composites exhibited unsatisfactory improvement due to structure defects or high interfacial thermal resistance between CNTs and polymer. Moreover, it is hard to disperse and align CNTs in polymer matrix. VACNTs become more popular due to the success on large-scale VACNTs growth by various CVD processes including common thermal CVD (TCVD) [112, 113], plasma-enhanced CVD (PECVD) [114, 115], floating catalyst CVD (FCCVD) [116, 117], hot filament CVD (HFCVD) [118], etc. Many researches have been done and reported large interface thermal resistance ranging from 1.7 to 25 mm^2KW^{-1} under various pressures using 3ω method, 1D reference bar method, photothermal method, photoacoustic technique, IR microscopy technique, thermal reflectance technique. The high interface thermal contact resistance inhibits the VCNTs TIMs real-life applications because of the weak interface of CNTs and mating substrate. The CNTs are not in real contact with the mating substrate surface since there is no chemical bonding at the interface.

Various VACNT bonding/transfer processes have been developed to address the challenge of modifying the CNT/mating substrate interface to effectively reduce the thermal contact resistance and simultaneously improve the interfacial adhesion. Zhu et al. proposed a solder transfer methodology for VACNTs [119]. By printing a thin layer of solder paste on substrate surface, reflowing, solidifying, and polishing the solder layer, VACNTs are able to form electrical and mechanical connections. This process overcomes the serious obstacles of integrating VACNTs into electronic packaging at low process temperatures and improved adhesion of VACNTs to substrates. Kordás et al. demonstrated a scalable assembly process similar to the above technique to fabricate VACNT microfins onto the chip with AuSn [120]. ~30 Wcm^{-2} and ~100 Wcm^{-2} more power at 100 °C from a hot chip have been found to dissipate from the CNT fin structure for the cases of natural and forced convections, respectively. The cooling performance of VACNTs makes them good candidates for on-chip thermal management applications. Kim et al. used a ceramic cement and filled the cement with Au particles to enhance electrical conductance of VACNTs on Si substrates [121]. Sunden et al. reported a microwave-assisted transfer of VACNTs onto thermoplastic polymer substrates [122]. Jiang et al. proposed a low-temperature transfer process using conductive polymer composites [123]. An ohmic contact was formed between a CNT film and a highly conductive polymer composite while a semiconductor joint was formed between a CNT film and a highly resistive polymer composite. Gan et al. reported a large-scale bonding of VACNTs to metal substrates with the assistance of high-frequency induction heating. This technique provides a potential approach to reproducible large-scale fabrication of VACNTs for various applications [115]. Recently, Lin developed a process combining in-situ functionalization and microwave curing and effectively enhanced the interface between carbon nanotubes and the epoxy matrix [124]. Effective medium theory has been used to analyze the interfacial thermal resistance between carbon nanotubes and polymer matrix, and that between graphite nanoplatelets and polymer matrix. Effective improvement of the interfacial thermal transport has been achieved by the interfacial bonding.

11.3.2 Carbon Nanotubes/Nanofibers (CNTs/CNFs) for Interconnects

With the development of downsizing, microwave switching based on diodes or transistors and fully passive devices like MEMS switches using electrostatic actuation of cantilevers or double clamped beams is approaching their inherent limitations of used technologies. The development of RF nano-electronics requires miniaturization and nano-scale interconnections. According to the International Technology Roadmap for Semiconductor, flip-chip bump pitches will shrink beyond 150 μm [125]. However, conventional metallic solder bumps have high diffusive and softening nature and are hard to downscale beyond 100 μm [126].

CNFs are a stack of graphene layers in a cone morphology which is more disordered than CNTs. The growth temperature of CNFs is lower than CNTs using PECVD [127], thus CNFs are considerably cheaper than CNT with a lower quality. They have been proposed to replace copper as vertical and horizontal interconnects in integrated circuits packaging [3, 127, 128]. Taking advantages from vertical vias and planar very low loss connections, CNFs/CNTs are particularly promising in developing nano-scale interconnections in 3D systems. Due to mechanical, thermal, and electrical properties, they are actively interested for applications in microelectronics, as well as in microwave devices. As compared with metal, CNFs/CNTs possess a higher current-carrying capacity, negligible skin depth effect, and no high-frequency current crowding issue because of their large kinetic inductance and negligible magnetic inductance [129].

The techniques of growing and patterning CNFs/CNTs are very important for its development in IC packaging. Few attempts have been done using localized growth of single CNF/CNT to build blocks in a sub-system, as well as patterning [130], planar-type devices [131], and aqueous solutions [132]. PECVD has been used widely to grow CNFs/CNTs by the introduction of electric field to achieve alignment as well as lower growth temperature [133]. The success of growing CNTs on Au opens up opportunities to apply vertically aligned CNT bundles in very high frequency domain.

The early studies of their application in RF electronics were performed on SWCNT [134–137] and CNF [138]. CNFs/CNTs bumps had been demonstrated by several groups as potential off-chip interconnects [139–141]. A low bundle resistance of 2.3 Ω (for a 100 μm diameter bump) and good mechanical flexibility have been demonstrated by Soga et al. [140]. Hermann et al. has reported a reliable CNFs/CNTs bump flip-chip interconnect over 2000 temperature cycles [139]. The application of CNFs/CNTs bumps for high-power amplifier application had also been demonstrated [141]. High thermal conductivity VACNTs flip-chip bump for high power dissipation has been demonstrated by Fujitsu Laboratory using selective growth [142]. Thanks to the small size of CNFs/CNTs (10 μm width for 15 μm height), the CNFs/CNTs bump for the flip-chip connecting the fine electrodes of the high power amplifiers (HPA) can reduce the ground inductance and maintain the same thermal conductivity as well (Fig. 11.2). A thermal conductivity of 1400 W/

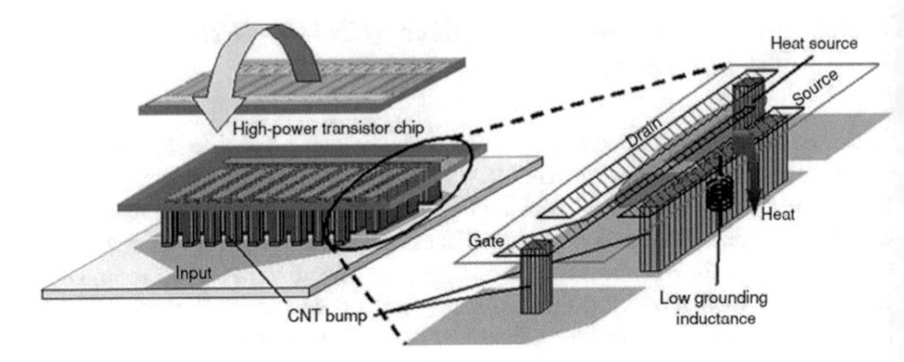

Fig. 11.3 Sketch of a flip-chip HPA with CNT bumps [142]

mK, which is three to four times higher than the one of Cu bumps, has been reported for CNFs/CNTs bumps. It has also been pointed out that the gain for 5 GHz and above has been improved at least 2 dB compared with face-up structure. It is believed that CNFs are able to be improved to be more closed to MWCNTs by applying some appropriate catalyst at a much lower cost compared with CNTs. It is thus be optimistic to envision applying CNFs/CNTs as interconnect via for chip of 32 nm node and beyond. However, the research of applying CNFs/CNTs for interconnects at microwave frequency is just in infancy (Fig. 11.3).

11.3.3 Graphene for Interconnects

Due to the superior electrical properties, graphene attracts great interest of research and application in transistors and interconnects. Same as CNTs, the miniature dimension is a big advantage for the application of graphene to interconnects and make it compatible with potential "beyond CMOS" transistors. Some other advantages of materials properties also include mechanical and thermal properties as discussed before. The application of graphene could resolve the electromigration and current density related issues of current metallic interconnects (typically Cu) [143]. Therefore active researches of graphene have been done for on-chip interconnects.

The resistivity was reported to be comparable to Cu interconnects in early studies [144, 145]. Later on, the combination of CNTs and graphene has also been investigated for 2.5D/3D through-silicon via (TSV) interconnects [143]. Using graphene as TSC filling and horizontal interconnects has also been proposed by several researchers [143, 146]. Large area flexible conducting graphene films have been demonstrated with less than 125 Ωm/sq sheet resistances using a CVD process [147]. It is also very potential to apply graphene for robust, low-cost anisotropic conducting films for interconnect applications, especially with transparent display. Promising electrical results have been reported to develop anisotropic conductive films (ACF) with graphene-encapsulated magnetic microspheres or graphene-coated polymer spheres [148]. Graphene has also been studied as a potential additive in solders. High thermal and electrical conductance has been demonstrated by incorporate exfoliated graphene oxide flakes into indium or indium-gallium alloy to form a matrix composition [149].

11.3.4 Graphene for Heat Spreading Layers

Thermal management has become very challenge and bottleneck in the state-of-the-art electronics due to rapid transistor scaling-down and increase of power densities [150, 151]. Efficient heat removal has become a critical issue for the performance and reliability of modern electronic, especially for RF/microwave applications. Development of the next generations of integrated circuits (ICs), high-frequency high-power density communication devices, makes the thermal management requirements extremely severe [97, 150–156]. The emerging 2.5D/3D integration of RF/microwave power amplifiers (based on Si- LDMOS, GaAs MESFETs, SiC MESFETs and GaN HEMTs technologies) with Si microprocessors and memory devices in consumer and military electronics requires high heat dissipation capability to get rid of hot spots on chip. Due to non-uniform heat generation and heat dissipation, micro-nano-scale hot spots are generated in high-power density electronics. The hot spots issue may result in performance degradation and reliability issues [157, 158]. Efficient thermal removal from hot spots area is very important for these high-power density devices such as SiC or GaN field-effect transistors (FETs). Micro-nano-scale heat spread would be a possible solution for hot spots cooling. However, thermal conductivity of metal films rapidly decreases with the film thickness [159–161] due to rough surface scattering, phonon grain-boundary scattering, and phonon confinement effects [162–169]. For many technologically important metals, e.g., aluminum, copper, or gold, the thermal conductivity of the 100 nm thick metal film is only ~20% of the thermal conductivity of bulk metal. Take the advantages of large thermal conductivity and inherent 2D nature, graphene is believed to be a good candidate material for heat spreading layers providing an efficient thermal conduction to resolve the hot spots issue without the complex on-chip active cooling system, such as microfluidic cooling or thin film thermoelectrics.

The first experimental feasibility study of graphene lateral heat spreaders for electronic devices was demonstrated on GaN-based HEMTs [170]. GaN-based wide bandgap III-N heterostructures have big advantages of electrical properties such as high breakdown voltage, high carrier density, high carrier mobility, and high carrier saturation velocity due to 2DEG (two-dimensional electron gas) forming at their heterojunction. Therefore GaN-based HEMTs are attractive for high-frequency, high-power, and high-temperature applications in consumer and military electronics. Amplifiers fabricated using AlGaN/GaN HFETs have produced RF power over a wide frequency range up to several hundred watts and several hundred GHz, that is an order of magnitude larger than GaAs or InP based power or RF devices.

Commercial AlGaN/GaN heterostructure field-effect transistors (HFETs) emerged in 2005 and have developed rapidly since then. However, there is a big problem of huge amount of heat generation during these devices' operation. Severe performance degradation and reliability issues rise up with self-overheating. Drain current, gain, output power and gate leakage increase and device lifetime decrease exponentially as operation temperature goes up. Extreme heat dissipation is in demand. The application of few-lay-graphene (FLG) heat spreader has been demonstrated to substantially improve the local thermal management of AlGaN/GaN HFETs [170]. In this research, highly oriented pyrolytic graphite (HOPG) has been used to exfoliate FLG, and polymethyl methacrylate (PMMA) membrane has been used as supporting materials to transfer FLG to AlGaN/GaN devices. The method provides a fast transfer with the accuracy of spatial alignment around 1–2 μm. However, this method cannot be applied in semiconductor industry due to the limitation of exfoliation. The graphene flake size, shape, thickness, and throughput are all limited and cannot be well controlled. The growth of large size, high quality, and low-price graphene will promote the practical applications of graphene heat spreaders. Fast progress of graphene growth by CVD method [147, 171, 172] and other techniques can make this feasible in the near future.

Several other experiment using graphene-based heat spreader has been reported to reduce hot spots on MOSFET devices on SOI substrates [173] as well as GaN HEMT devices [170] with potential increase of yield up to an order of magnitude. Graphene can realize lateral heat spreading due to its 2D structure. In a 3D stack, lateral heat spreading enables utilizing a larger layer-area for interlayer heat conduction so as to improve the overall package thermal conductivity.

11.3.5 Graphene for Other Applications of IC Packaging

Graphene-based materials like graphene flakes or graphene nanoplatelets (GNP) are widely used as fillers to form composite in polymeric matrix. These composites are being applied as conductive inks, thermal interface materials (TIMs), barrier layers, shielding layers, encapsulants, and electrically conductive adhesives. Enhancements in mechanical strength, electrical and thermal conductivity, and gas barrier properties have been reported as a result of the application of graphene-based composite [174]. An increase factor of 2–8 in elastic modulus and yield strength has been reported with <3% volume fraction of graphene-based fillers in common polymeric matrix systems such as PDMS, PVA (poly (vinyl alcohol)), and PCL (polycaprolactone) [174]. Similarly, up to 2300% enhancement in thermal conductivity of graphene-based TIM has been achieved at 10% graphene loading, outperforming conventional metallic fillers as well as CNT fillers in similar matrix systems [174]. Other research about the application of graphene for barrier layer has demonstrated as much as 90% gas permeation reduction at only 3 wt% of graphene in oriented parallel polymer films [174]. Thus, many potential applications to enhance functional properties in microelectronics packaging can be developed using graphene-based composite materials due to its excellent intrinsic properties and their 2D nature (as opposed to CNTs and other nano-fillers).

11.4 Challenges

Many challenges need to be overcome in applications utilizing both CNTs and graphene. The challenges discussed below pertain to the fabrication processes, as well as to the inherent structure of the implementations. As pointed out before, the qualities of CNT and graphene play a very important role on their performance. There is a big challenge to grow these materials with perfect quality, low cost and compatible with process. PECVD is good at growing and patterning CNTs/graphene directly onto device substrate at relatively low temperature and compatible with current process, but the quality is not as good as TCVD. But TCVD requires temperatures as high as 1000 °C and subsequently transferred to the device substrate [175]. The high temperature is not compatible with current CMOS process and CNTs/graphene need to be grown on other substrates and transfer to device wafers later. However, the transferring process is hard to promise not to introduce contaminations of defects. In most cases, after transferring process, the performance is much lower than its intrinsic properties due to poor chemical interface bonding and tip-to-tip contact. Simulation results suggest that the spacing and length of the CNT connections and the precise nature of CNT-graphene nodes need careful consideration [176]. Both CNTs/CNFs and graphene implementations are inhibited by the lack of a viable fabrication process. Challenges also rise when CNTs/graphene are used as fillers to form nanocomposite. As discussed earlier, such materials can be used as inks, TIMs, barrier coatings, encapsulants, etc. In this context, top-down approaches to manufacture graphene-based fillers (GNP, flakes, etc.) are particularly attractive. While initial results in this field have been quite promising, there is still considerable room for improvement and optimization. For example, the most attractive method of large-scale production of graphene-based nanocomposites is through the oxidation of graphite and thermal or chemical reduction. But this method can also adversely modify the mechanical stiffness, electrical transport, and thermal properties of the graphene fillers. What's more, wrinkles, bending, misalignment, and poor interface contacts may make the nanocomposite less stiff, more permeable and reduce thermal and electrical conductivity. Another big concern is the cost. How to grow high quality CNTs/graphene while keep the cost at a low level comparable with metallic packaging materials is a challenging topic need to be fixed by researchers/engineers. Alternative cost-effective routes to produce CNTs and graphene need to be sought to preserve their intrinsic properties in the nanocomposite system. These would involve eliminating the structural defects, unnecessary functionalization, post-treatments or some innovative ideas of growth and process.

11.5 Summary

CNTs and graphene possess excellent mechanical, electrical, and thermal properties. They are proposed to be used in RF electronic packaging due their intrinsic properties. This chapter reviewed the state of the art of three particular applications: (i) CNT for TIMs, (ii) CNT for interconnect, (iii) Graphene for interconnects, (iv) Graphene for heat spreading layer, and (v) Graphene for other applications of IC packaging. The mechanical, electrical, and thermal properties of CNTs and graphene are also briefly introduced. Although the applications of CNTs and graphene in RF packaging are very promising due to these advanced properties, there are also many challenges for practical application including deposition method and patterning compatibility with conventional devices fabrication process, post-deposition treatment and transfer effect on defect and contamination level, poor interface bonding and tip-tip contact, and low efficiency and high production cost. Further improvement can be probably made to improve crystal quality, process capability, productivity, and filler (CNTs or graphene) /matrix interface contact quality to approaching the good intrinsic properties of CNTs/graphene.

References

1. Moore, GE. 1975. Progress in digital integrated electronics. *Electron Devices Meeting, 1975 International*, (IEEE), 11–13.
2. Steinlesberger, G., et al. 2002. Electrical assessment of copper damascene interconnects down to sub-50 nm feature sizes. *Microelectronic Engineering* 64(1–4): 409–416.
3. Kreupl, F., et al. 2002. Carbon nanotubes in interconnect applications. *Microelectronic Engineering* 64(1–4): 399–408.
4. Li, J., et al. 2003. Bottom-up approach for carbon nanotube interconnects. *Applied Physics Letters* 82(15): 2491–2493.
5. Naeemi, A., R. Sarvari, and J.D. Meindl. 2005. Performance comparison between carbon nanotube and copper interconnects for gigascale integration (GSI). *IEEE Electron Device Letters* 26(2): 84–86.
6. Mizuhisa, N., H. Masahiro, K. Akio, and A. Yuji. 2004. Simultaneous formation of multiwall carbon nanotubes and their end-bonded ohmic contacts to Ti electrodes for future ULSI interconnects. *Japanese Journal of Applied Physics* 43(4S): 1856.
7. Naeemi, A., and J.D. Meindl. 2005. Impact of electron-phonon scattering on the performance of carbon nanotube interconnects for GSI. *IEEE Electron Device Letters* 26(7): 476–478.
8. Myounggu, P., et al. 2006. Effects of a carbon nanotube layer on electrical contact resistance between copper substrates. *Nanotechnology* 17(9): 2294.
9. Nieuwoudt, A., and Y. Massoud. 2006. Understanding the impact of inductance in carbon nanotube bundles for VLSI interconnect using scalable modeling techniques. *IEEE Transactions on Nanotechnology* 5(6): 758–765.
10. Dresselhaus, M., G. Dresselhaus, and R. Saito. 1995. Physics of carbon nanotubes. *Carbon* 33(7): 883–891.
11. Lu, X., and Z. Hu. 2012. Mechanical property evaluation of single-walled carbon nanotubes by finite element modeling. *Composites Part B: Engineering* 43(4): 1902–1913.
12. Yakobson, B.I., C.J. Brabec, and J. Bernholc. 1996. Nanomechanics of carbon tubes: instabilities beyond linear response. *Physical Review Letters* 76(14): 2511–2514.

13. Qiang, L., A. Marino, and H. Rui. 2009. Elastic bending modulus of monolayer graphene. *Journal of Physics D: Applied Physics* 42(10): 102002.
14. Liew, K.M., X.Q. He, and C.H. Wong. 2004. On the study of elastic and plastic properties of multi-walled carbon nanotubes under axial tension using molecular dynamics simulation. *Acta Materialia* 52(9): 2521–2527.
15. Lourie, O., D.M. Cox, and H.D. Wagner. 1998. Buckling and collapse of embedded carbon nanotubes. *Physical Review Letters* 81(8): 1638–1641.
16. Zhang, C.-L., and H.-S. Shen. 2006. Buckling and postbuckling analysis of single-walled carbon nanotubes in thermal environments via molecular dynamics simulation. *Carbon* 44(13): 2608–2616.
17. Lee, C., X. Wei, J.W. Kysar, and J. Hone. 2008. Measurement of the elastic properties and intrinsic strength of monolayer graphene. *Science* 321(5887): 385.
18. Ricardo, F., A.D. Pablo, P. Helena, G. Cecilia, and W.M. Álvaro. 2009. Mechanical properties of graphene nanoribbons. *Journal of Physics: Condensed Matter* 21(28): 285304.
19. Liu F, Ming P, and Li J (2007) *Ab initio* calculation of ideal strength and phonon instability of graphene under tension. *Physical Review B* 76(6): 064120.
20. Hernández, E., C. Goze, P. Bernier, and A. Rubio. 1998. Elastic properties of C and $B_xC_yN_z$ composite nanotubes. *Physical Review Letters* 80(20): 4502–4505.
21. Ávila, A.F., and G.S.R. Lacerda. 2008. Molecular mechanics applied to single-walled carbon nanotubes. *Materials Research* 11: 325–333.
22. Guanghua, G., Ç. Tahir, and A.G. William III. 1998. Energetics, structure, mechanical and vibrational properties of single-walled carbon nanotubes. *Nanotechnology* 9(3): 184.
23. Li, C., and T.-W. Chou. 2003. Elastic moduli of multi-walled carbon nanotubes and the effect of van der Waals forces. *Composites Science and Technology* 63(11): 1517–1524.
24. Treacy, M.M.J., T.W. Ebbesen, and J.M. Gibson. 1996. Exceptionally high Young's modulus observed for individual carbon nanotubes. *Nature* 381(6584): 678–680.
25. Wong, E.W., P.E. Sheehan, and C.M. Lieber. 1997. Nanobeam mechanics: elasticity, strength, and toughness of nanorods and nanotubes. *Science* 277(5334): 1971.
26. Lu, J.P. 1997. Elastic properties of carbon nanotubes and nanoropes. *Physical Review Letters* 79(7): 1297–1300.
27. Krishnan, A., E. Dujardin, T.W. Ebbesen, P.N. Yianilos, and M.M.J. Treacy. 1998. Young's modulus of single-walled nanotubes. *Physical Review B* 58(20): 14013–14019.
28. Salvetat, J.-P., et al. 1999. Mechanical properties of carbon nanotubes. *Applied Physics A* 69(3): 255–260.
29. Yu, M.-F., B.S. Files, S. Arepalli, and R.S. Ruoff. 2000. Tensile loading of ropes of single wall carbon nanotubes and their mechanical properties. *Physical Review Letters* 84(24): 5552–5555.
30. Tombler, T.W., et al. 2000. Reversible electromechanical characteristics of carbon nanotubes under local-probe manipulation. *Nature* 405(6788): 769–772.
31. Demczyk, B.G., et al. 2002. Direct mechanical measurement of the tensile strength and elastic modulus of multiwalled carbon nanotubes. *Materials Science and Engineering A* 334(1–2): 173–178.
32. Xiao, J.R., B.A. Gama, and J.W. Gillespie Jr. 2005. An analytical molecular structural mechanics model for the mechanical properties of carbon nanotubes. *International Journal of Solids and Structures* 42(11–12): 3075–3092.
33. Kirtania, S., and D. Chakraborty. 2007. Finite element based characterization of carbon nanotubes. *Journal of Reinforced Plastics and Composites* 26(15): 1557–1570.
34. Ding, W., et al. 2007. Modulus, fracture strength, and brittle vs plastic response of the outer shell of Arc-grown multi-walled carbon nanotubes. *Experimental Mechanics* 47(1): 25–36.
35. Ji-nan, L., and C. Hai-bo. 2008. Analysis of single-walled carbon nanotubes using a chemical bond element model. *Chinese Journal of Chemical Physics* 21(4): 353.
36. Huang, M. 2009. Studies of mechanically deformed single walled carbon nanotubes and graphene by optical spectroscopy. PhD, Columbia University.

37. Rossi, M., and M. Meo. 2009. On the estimation of mechanical properties of single-walled carbon nanotubes by using a molecular-mechanics based FE approach. *Composites Science and Technology* 69(9): 1394–1398.
38. Meo, M., and M. Rossi. 2006. Prediction of Young's modulus of single wall carbon nanotubes by molecular-mechanics based finite element modelling. *Composites Science and Technology* 66(11–12): 1597–1605.
39. Blakslee, O.L., D.G. Proctor, E.J. Seldin, G.B. Spence, and T. Weng. 1970. Elastic constants of compression-annealed pyrolytic graphite. *Journal of Applied Physics* 41(8): 3373–3382.
40. Bunch, J.S., et al. 2008. Impermeable atomic membranes from graphene sheets. *Nano Letters* 8(8): 2458–2462.
41. Shokrieh, M.M., and R. Rafiee. 2010. Prediction of Young's modulus of graphene sheets and carbon nanotubes using nanoscale continuum mechanics approach. *Materials & Design* 31(2): 790–795.
42. Odom, T.W., J.-L. Huang, P. Kim, and C.M. Lieber. 1998. Atomic structure and electronic properties of single-walled carbon nanotubes. *Nature* 391(6662): 62–64.
43. Wilder, J.W.G., L.C. Venema, A.G. Rinzler, R.E. Smalley, and C. Dekker. 1998. Electronic structure of atomically resolved carbon nanotubes. *Nature* 391(6662): 59–62.
44. Peierls, R.E. 1955. *Quantum Theory of Solids*. Oxford: Clarendon Press.
45. Yao, Z., C.L. Kane, and C. Dekker. 2000. High-field electrical transport in single-wall carbon nanotubes. *Physical Review Letters* 84(13): 2941–2944.
46. Wei, B.Q., R. Vajtai, and P.M. Ajayan. 2001. Reliability and current carrying capacity of carbon nanotubes. *Applied Physics Letters* 79(8): 1172–1174.
47. Naeemi, A., and J.D. Meindl. 2006. Compact physical models for multiwall carbon-nanotube interconnects. *IEEE Electron Device Letters* 27(5): 338–340.
48. Matsumura, H., and T. Ando. 2001. Conductance of carbon nanotubes with a Stone–Wales defect. *Journal of the Physical Society of Japan* 70(9): 2657–2665.
49. Watts, P.C.P., W.-K. Hsu, H.W. Kroto, and D.R.M. Walton. 2003. Are bulk defective carbon nanotubes less electrically conducting? *Nano Letters* 3(4): 549–553.
50. Reyes, S.A., A. Struck, and S. Eggert. 2009. Lattice defects and boundaries in conducting carbon nanotubes. *Physical Review B* 80(7): 075115.
51. Yoon, Y.-G., M.S.C. Mazzoni, H.J. Choi, J. Ihm, and S.G. Louie. 2001. Structural deformation and intertube conductance of crossed carbon nanotube junctions. *Physical Review Letters* 86(4): 688–691.
52. Buldum, A., and J.P. Lu. 2001. Contact resistance between carbon nanotubes. *Physical Review B* 63(16): 161403.
53. Tersoff, J. 1999. Contact resistance of carbon nanotubes. *Applied Physics Letters* 74(15): 2122–2124.
54. Lee, J.O., et al. 2000. Formation of low-resistance ohmic contacts between carbon nanotube and metal electrodes by a rapid thermal annealing method. *Journal of Physics* 33((16)): 1953–1956.
55. Stetter, A.V.J., and C.H. Back. Conductivity of multiwall carbon nanotubes: role of multiple shells and defects. *Physical Review B* 82((11)): 115451–1151e5.
56. Novoselov, K.S., et al. 2004. Electric field effect in atomically thin carbon films. *Science* 306(5696): 666.
57. Chen, J.-H., C. Jang, S. Xiao, M. Ishigami, and M.S. Fuhrer. 2008. Intrinsic and extrinsic performance limits of graphene devices on SiO_2. *Nature Nanotechnology* 3(4): 206–209.
58. Baughman RH, Zakhidov AA, and de Heer WA (2002) Carbon nanotubes—The route toward applications Science297(5582):787.
59. Pop, E., D. Mann, Q. Wang, K. Goodson, and H. Dai. 2006. Thermal conductance of an individual single-wall carbon nanotube above room temperature. *Nano Letters* 6(1): 96–100.
60. Fujii, M., et al. 2005. Measuring the thermal conductivity of a single carbon nanotube. *Physical Review Letters* 95(6): 065502.
61. Ren, C., W. Zhang, Z. Xu, Z. Zhu, and P. Huai. 2010. Thermal conductivity of single-walled carbon nanotubes under axial stress. *The Journal of Physical Chemistry C* 114(13): 5786–5791.

62. Guthy, C., F. Du, S. Brand, K.I. Winey, and J.E. Fischer. 2007. Thermal conductivity of single-walled carbon nanotube/PMMA nanocomposites. *Journal of Heat Transfer* 129(8): 1096–1099.
63. Hone, J., M. Whitney, C. Piskoti, and A. Zettl. 1999. Thermal conductivity of single-walled carbon nanotubes. *Physical Review B* 59(4): R2514–R2516.
64. Savin, A.V., B. Hu, and Y.S. Kivshar. 2009. Thermal conductivity of single-walled carbon nanotubes. *Physical Review B* 80(19): 195423.
65. Mingo, N., and D.A. Broido. 2005. Carbon nanotube ballistic thermal conductance and its limits. *Physical Review Letters* 95(9): 096105.
66. Mohamed, A.O., and S. Deepak. 2001. Temperature dependence of the thermal conductivity of single-wall carbon nanotubes. *Nanotechnology* 12(1): 21.
67. Yu, C., L. Shi, Z. Yao, D. Li, and A. Majumdar. 2005. Thermal conductance and thermopower of an individual single-wall carbon nanotube. *Nano Letters* 5(9): 1842–1846.
68. Kwon, Y.-K., and P Kim. 2006. Unusually high thermal conductivity in carbon nanotubes. In *High Thermal Conductivity Materials*, eds. SL Shindé and JS Goela, 227–265. New York: Springer.
69. Berber, S., Y.-K. Kwon, and D. Tománek. 2000. Unusually high thermal conductivity of carbon nanotubes. *Physical Review Letters* 84(20): 4613–4616.
70. Jianwei, C., Ç. Tahir, and A.G. William III. 2000. Thermal conductivity of carbon nanotubes. *Nanotechnology* 11(2): 65.
71. Choi, T.-Y., D. Poulikakos, J. Tharian, and U. Sennhauser. 2006. Measurement of the thermal conductivity of individual carbon nanotubes by the four-point three-ω method. *Nano Letters* 6(8): 1589–1593.
72. Mingo, N., and D.A. Broido. 2005. Length dependence of carbon nanotube thermal conductivity and the "problem of long waves". *Nano Letters* 5(7): 1221–1225.
73. Takahiro, Y., N. Yoshiki, and W. Kazuyuki. 2007. Control of electron- and phonon-derived thermal conductances in carbon nanotubes. *New Journal of Physics* 9(8): 245.
74. Wei, Z., et al. 2004. Chirality dependence of the thermal conductivity of carbon nanotubes. *Nanotechnology* 15(8): 936.
75. Kim, P., L. Shi, A. Majumdar, and P.L. McEuen. 2001. Thermal transport measurements of individual multiwalled nanotubes. *Physical Review Letters* 87(21): 215502.
76. Cao, J.X., X.H. Yan, Y. Xiao, and J.W. Ding. 2004. Thermal conductivity of zigzag single-walled carbon nanotubes: role of the umklapp process. *Physical Review B* 69(7): 073407.
77. Maruyama, S. 2003. A Molecular dynamics simulation of heat conduction of a finite length single-walled carbon nanotube. *Microscale Thermophysical Engineering* 7(1): 41–50.
78. Padgett, C.W., and D.W. Brenner. 2004. Influence of chemisorption on the thermal conductivity of single-wall carbon nanotubes. *Nano Letters* 4(6): 1051–1053.
79. Tang, K., et al. 2013. Molecular dynamics simulation on thermal conductivity of single-walled carbon nanotubes. In *Electronic Packaging Technology (ICEPT), 2013 14th International Conference on*, 583–586.
80. Guo, Z.-x., and Gong X-g. 2009. Molecular dynamics studies on the thermal conductivity of single-walled carbon nanotubes. *Frontiers of Physics in China* 4(3): 389–392.
81. Pettes, M.T., and L. Shi. 2009. Thermal and structural characterizations of individual single-, double-, and multi-walled carbon nanotubes. *Advanced Functional Materials* 19(24): 3918–3925.
82. Yamamoto, T., and K. Watanabe. 2006. Nonequilibrium Green's function approach to phonon transport in defective carbon nanotubes. *Physical Review Letters* 96(25): 255503.
83. Wang, Z.L., et al. 2007. Length-dependent thermal conductivity of an individual single-wall carbon nanotube. *Applied Physics Letters* 91(12): 123119.
84. Mizel, A., et al. 1999. Analysis of the low-temperature specific heat of multiwalled carbon nanotubes and carbon nanotube ropes. *Physical Review B* 60(5): 3264–3270.
85. Yang, D.J., et al. 2002. Thermal conductivity of multiwalled carbon nanotubes. *Physical Review B* 66(16): 165440.
86. Hone, J., B. Batlogg, Z. Benes, A.T. Johnson, and J.E. Fischer. 2000. Quantized phonon spectrum of single-wall carbon nanotubes. *Science* 289(5485): 1730.

87. Balandin, A.A., et al. 2008. Superior thermal conductivity of single-layer graphene. *Nano Letters* 8(3): 902–907.
88. Cai, W., et al. 2010. Thermal transport in suspended and supported monolayer graphene grown by chemical vapor deposition. *Nano Letters* 10(5): 1645–1651.
89. Faugeras, C., et al. 2010. Thermal conductivity of graphene in corbino membrane geometry. *ACS Nano* 4(4): 1889–1892.
90. Xu X, et al. 2014. Length-dependent thermal conductivity in suspended single-layer graphene. *Nature Communications* 5. Article number:3689.
91. Lee, J.-U., D. Yoon, H. Kim, S.W. Lee, and H. Cheong. 2011. Thermal conductivity of suspended pristine graphene measured by Raman spectroscopy. *Physical Review B* 83(8): 081419.
92. Seol, J.H., et al. 2010. Two-dimensional phonon transport in supported graphene. *Science* 328(5975): 213.
93. Klemens, P.G. 2001. Theory of thermal conduction in thin ceramic films. *International Journal of Thermophysics* 22(1): 265–275.
94. Pettes, M.T., I. Jo, Z. Yao, and L. Shi. 2011. Influence of polymeric residue on the thermal conductivity of suspended bilayer graphene. *Nano Letters* 11(3): 1195–1200.
95. Xu, J., and T.S. Fisher. 2006. Enhancement of thermal interface materials with carbon nanotube arrays. *International Journal of Heat and Mass Transfer* 49(9–10): 1658–1666.
96. Gwinn, J.P., and R.L. Webb. 2003. Performance and testing of thermal interface materials. *Microelectronics Journal* 34(3): 215–222.
97. Prasher, R. 2006. Thermal interface materials: historical perspective, status, and future directions. *Proceedings of the IEEE* 94(8): 1571–1586.
98. Prasher, R.S. 2001. Surface chemistry and characteristics based model for the thermal contact resistance of fluidic interstitial thermal interface materials. *Journal of Heat Transfer* 123(5): 969–975.
99. Yu, A., P. Ramesh, M.E. Itkis, E. Bekyarova, and R.C. Haddon. 2007. Graphite nanoplatelet–epoxy composite thermal interface materials. *The Journal of Physical Chemistry C* 111(21): 7565–7569.
100. Aoyagi, Y., C.-K. Leong, and D.D.L. Chung. 2006. Polyol-based phase-change thermal interface materials. *Journal of Electronic Materials* 35(3): 416–424.
101. Chung, D.D.L. 2001. Thermal interface materials. *Journal of Materials Engineering and Performance* 10(1): 56–59.
102. Howe, T.A., C.-K. Leong, and D.D.L. Chung. 2006. Comparative evaluation of thermal interface materials for improving the thermal contact between an operating computer microprocessor and its heat sink. *Journal of Electronic Materials* 35(8): 1628–1635.
103. Liu, Z., and D.D.L. Chung. 2001. Calorimetric evaluation of phase change materials for use as thermal interface materials. *Thermochimica Acta* 366(2): 135–147.
104. Maguire, L., M. Behnia, and G. Morrison. 2005. Systematic evaluation of thermal interface materials—a case study in high power amplifier design. *Microelectronics Reliability* 45(3–4): 711–725.
105. Prasher, R.S. 2004. Rheology based modeling and design of particle laden polymeric thermal interface materials. In *Thermal and Thermomechanical Phenomena in Electronic Systems, 2004. ITHERM'04. The Ninth Intersociety Conference on*, Vol. 31, 36–44.
106. Vishal, S., T. Siegmund, and S.V. Garimella. 2004. Optimization of thermal interface materials for electronics cooling applications. *IEEE Transactions on Components and Packaging Technologies* 27(2): 244–252.
107. Pour Shahid Saeed Abadi, P., C.-K. Leong, and D.D.L. Chung. 2008. Factors that govern the performance of thermal interface materials. *Journal of Electronic Materials* 38(1): 175–192.
108. Aoyagi, Y., and D.D.L. Chung. 2008. Antioxidant-based phase-change thermal interface materials with high thermal stability. *Journal of Electronic Materials* 37(4): 448–461.
109. De Mey, G., et al. 2009. Influence of interface materials on the thermal impedance of electronic packages. *International Communications in Heat and Mass Transfer* 36(3): 210–212.

110. Lin, C., and D.D.L. Chung. 2009. Graphite nanoplatelet pastes vs. carbon black pastes as thermal interface materials. *Carbon* 47(1): 295–305.
111. Liu, X., Y. Zhang, A.M. Cassell, and B.A. Cruden. 2008. Implications of catalyst control for carbon nanotube based thermal interface materials. *Journal of Applied Physics* 104(8): 084310.
112. Hata, K., et al. 2004. Water-assisted highly efficient synthesis of impurity-free single-walled carbon nanotubes. *Science* 306(5700): 1362.
113. Zhu, L., Y. Xiu, D.W. Hess, and C.-P. Wong. 2005. Aligned carbon nanotube stacks by water-assisted selective etching. *Nano Letters* 5(12): 2641–2645.
114. Qu, L., and L. Dai. 2007. Gecko-foot-mimetic aligned single-walled carbon nanotube dry adhesives with unique electrical and thermal properties. *Advanced Materials* 19(22): 3844–3849.
115. Hofmann, S., C. Ducati, J. Robertson, and B. Kleinsorge. 2003. Low-temperature growth of carbon nanotubes by plasma-enhanced chemical vapor deposition. *Applied Physics Letters* 83(1): 135–137.
116. Nikolaev, P., et al. 1999. Gas-phase catalytic growth of single-walled carbon nanotubes from carbon monoxide. *Chemical Physics Letters* 313(1–2): 91–97.
117. Andrews, R., et al. 1999. Continuous production of aligned carbon nanotubes: a step closer to commercial realization. *Chemical Physics Letters* 303(5–6): 467–474.
118. Xu, Y.-Q., et al. 2006. Vertical array growth of small diameter single-walled carbon nanotubes. *Journal of the American Chemical Society* 128(20): 6560–6561.
119. Zhu, L., Y. Sun, D.W. Hess, and C.-P. Wong. 2006. Well-aligned open-ended carbon nanotube architectures: an approach for device assembly. *Nano Letters* 6(2): 243–247.
120. Kordás, K., et al. 2007. Chip cooling with integrated carbon nanotube microfin architectures. *Applied Physics Letters* 90(12): 123105.
121. Kim, M.J., et al. 2006. Efficient transfer of a VA-SWNT film by a flip-over technique. *Journal of the American Chemical Society* 128(29): 9312–9313.
122. Sunden, E., J.K. Moon, C.P. Wong, W.P. King, and S. Graham. 2006. Microwave assisted patterning of vertically aligned carbon nanotubes onto polymer substrates. *Journal of Vacuum Science & Technology B* 24(4): 1947–1950.
123. Hongjin, J., Z. Lingbo, M. Kyoung-sik, and C.P. Wong. 2007. Low temperature carbon nanotube film transfer via conductive polymer composites. *Nanotechnology* 18(12): 125203.
124. Lin W and Wong CP (2010) Applications of carbon nanomaterials as electrical interconnects and thermal interface materials. In Nano-Bio-Electronic, Photonic and MEMS Packaging, eds Wong CP, Moon K-S, and Li Y. 87–138. Boston: Springer.
125. Semiconductor Industry Association (SIA). Assembly and Packaging. In International Technology Roadmap for Semiconductors, 2009 Edition. http://www.itrs.net/Links/2009ITRS/2009Chapters_2009Tables/2009_Assembly.pdf
126. Tummala, R., C.P. Wong, and P.M. Raj. 2009. Nanopackaging research at Georgia Tech. *IEEE Nanotechnology Magazine* 3(4): 20–25.
127. Ngo, Q., et al. 2007. Structural and electrical characterization of carbon nanofibers for interconnect via applications. *IEEE Transactions on Nanotechnology* 6(6): 688–695.
128. Coiffic, J.C., et al. 2008. An application of carbon nanotubes for integrated circuit interconnects. 70370D-70370D-70310.
129. Li, H., C. Xu, N. Srivastava, and K. Banerjee. 2009. Carbon nanomaterials for next-generation interconnects and passives: physics, status, and prospects. *IEEE Transactions on Electron Devices* 56(9): 1799–1821.
130. Chae, J., X. Ho, J.A. Rogers, and K. Jain. 2008. Patterning of single walled carbon nanotubes using a low-fluence excimer laser photoablation process. *Applied Physics Letters* 92(17): 173115.
131. Pribat, D., et al. 2008. Novel approach to align carbon nanotubes for planar type devices. 70370 N–70370 N-70377.
132. Shoji, S., T. Roders, Z. Sekkat, and S. Kawata. 2008. Light-induced accumulation of single-wall carbon nanotubes dispersed in aqueous solution. 70370O-70370O-70376.

133. Chhowalla, M., et al. 2001. Growth process conditions of vertically aligned carbon nanotubes using plasma enhanced chemical vapor deposition. *Journal of Applied Physics* 90(10): 5308–5317.
134. Burke, P.J., Z. Yu, S. Li, and C. Rutherglen. 2005. Nanotube technology for microwave applications. In *IEEE MTT-S International Microwave Symposium Digest, 2005*, 4.
135. Moayed, N.N.A., U.A. Khan, M. Obol, S. Gupta, and M.N. Afsar (2007) Characterization of single- and multi-walled carbon nanotubes at microwave frequencies. In *2007 IEEE Instrumentation & Measurement Technology Conference IMTC 2007*, 1–4.
136. Plombon, J.J., K.P. O'Brien, F. Gstrein, V.M. Dubin, and Y. Jiao. 2007. High-frequency electrical properties of individual and bundled carbon nanotubes. *Applied Physics Letters* 90(6): 063106.
137. Min, Z., H. Xiao, P.C.H. Chan, L. Qi, and Z.K. Tang. 2006. Radio-frequency transmission properties of carbon nanotubes in a field-effect transistor configuration. *IEEE Electron Device Letters* 27(8): 668–670.
138. Madriz, F.R., et al. 2008. Measurements and circuit model of carbon nanofibers at microwave frequencies. In *2008 International Interconnect Technology Conference*, 138–140.
139. Hermann, S., B. Pahl, R. Ecke, S.E. Schulz, and T. Gessner. 2010. Carbon nanotubes for nanoscale low temperature flip chip connections. *Microelectronic Engineering* 87(3): 438–442.
140. Ikuo, S., et al. 2008. Carbon nanotube bumps for LSI interconnect. In *2008 58th Electronic Components and Technology Conference*, 1390–1394.
141. Iwai, T., et al. 2005. Thermal and source bumps utilizing carbon nanotubes for flip-chip high power amplifiers. In *IEEE International Electron Devices Meeting, 2005. IEDM Technical Digest*, 257–260.
142. Awano, V., T. Iwai, and V. Yuji. 2007. Carbon nanotube bumps for thermal and electric conduction in transistor. *Fujitsu Scientific & Technical Journal* 43(4): 508–515.
143. Nihei, M. 2012. CNT/Graphene Technologies for Advanced Interconnects and TSVs. In *SEMATECH Symposium Taiwan*.
144. Murali, R., Y. Yang, K. Brenner, T. Beck, and J.D. Meindl. 2009. Breakdown current density of graphene nanoribbons. *Applied Physics Letters* 94(24): 243114.
145. Murali, R., K. Brenner, Y. Yang, T. Beck, and J.D. Meindl. 2009. Resistivity of graphene nanoribbon interconnects. *IEEE Electron Device Letters* 30(6): 611–613.
146. Kawabata, A., T. Murakami, M. Nihei, and N. Yokoyama. 2013. Growth of dense, vertical and horizontal graphene and its thermal properties. *Japanese Journal of Applied Physics* 52((4S)): 04CB06.
147. Bae, S., et al. 2010. Roll-to-roll production of 30-inch graphene films for transparent electrodes. *Nature Nanotechnology* 5(8): 574–578.
148. Shen, J., Y. Zhu, K. Zhou, X. Yang, and C. Li. 2012. Tailored anisotropic magnetic conductive film assembled from graphene-encapsulated multifunctional magnetic composite microspheres. *Journal of Materials Chemistry* 22(2): 545–550.
149. Jagannadham, K. 2011. Thermal conductivity of indium–graphene and indium-gallium–graphene composites. *Journal of Electronic Materials* 40(1): 25–34.
150. Garimella, S.V., et al. 2008. Thermal challenges in next-generation electronic systems. *IEEE Transactions on Components and Packaging Technologies* 31(4): 801–815.
151. Balandin, A.A. 2009. Chill out. *IEEE Spectrum* 46(10): 34–39.
152. Meneghini, M., L.-R. Trevisanello, G. Meneghesso, and E. Zanoni. 2008. A review on the reliability of GaN-based LEDs. *IEEE Transactions on Device and Materials Reliability* 8(2): 323–331.
153. Meneghesso, G., et al. 2008. Reliability of GaN high-electron-mobility transistors: state of the art and perspectives. *IEEE Transactions on Device and Materials Reliability* 8(2): 332–343.
154. Christensen, A., and S. Graham. 2009. Thermal effects in packaging high power light emitting diode arrays. *Applied Thermal Engineering* 29(2): 364–371.

155. Puttaswamy, K., and G.H. Loh. 2006. Thermal analysis of a 3D die-stacked high-performance microprocessor. In *Proceedings of the 16th ACM Great Lakes Symposium on VLSI, (ACM)*, 19–24.
156. Sarvar, F., D.C. Whalley, and P.P. Conway. 2006. Thermal interface materials-a review of the state of the art. In *2006 1st Electronic Systemintegration Technology Conference*, 1292–1302. IEEE.
157. Sarua, A., et al. 2006. Integrated micro-Raman/infrared thermography probe for monitoring of self-heating in AlGaN/GaN transistor structures. *IEEE Transactions on Electron Devices* 53(10): 2438–2447.
158. Turin, V.O., and A.A. Balandin. 2006. Electrothermal simulation of the self-heating effects in GaN-based field-effect transistors. *Journal of Applied Physics* 100(5): 054501.
159. Langer, G., J. Hartmann, and M. Reichling. 1997. Thermal conductivity of thin metallic films measured by photothermal profile analysis. *Review of Scientific Instruments* 68(3): 1510–1513.
160. Chen, G., and P. Hui. 1999. Thermal conductivities of evaporated gold films on silicon and glass. *Applied Physics Letters* 74(20): 2942–2944.
161. Nath, P., and K. Chopra. 1974. Thermal conductivity of copper films. *Thin Solid Films* 20(1): 53–62.
162. Nika, D.L., A.I. Cocemasov, D.V. Crismari, and A.A. Balandin. 2013. Thermal conductivity inhibition in phonon engineered core-shell cross-section modulated Si/Ge nanowires. *Applied Physics Letters* 102(21): 213109.
163. Nika, D., et al. 2012. Suppression of phonon heat conduction in cross-section-modulated nanowires. *Physical Review B* 85(20): 205439.
164. ———. 2011. Reduction of lattice thermal conductivity in one-dimensional quantum-dot superlattices due to phonon filtering. *Physical Review B* 84(16): 165415.
165. Martin, P., Z. Aksamija, E. Pop, and U. Ravaioli. 2009. Impact of phonon-surface roughness scattering on thermal conductivity of thin Si nanowires. *Physical Review Letters* 102(12): 125503.
166. Hochbaum, A.I., et al. 2008. Enhanced thermoelectric performance of rough silicon nanowires. *Nature* 451(7175): 163–167.
167. Li, D., et al. 2003. Thermal conductivity of individual silicon nanowires. *Applied Physics Letters* 83(14): 2934–2936.
168. Liu, W., and M. Asheghi. 2006. Thermal conductivity measurements of ultra-thin single crystal silicon layers. *Journal of Heat Transfer* 128(1): 75–83.
169. Zincenco, N., D. Nika, E. Pokatilov, and A. Balandin. 2007. Acoustic phonon engineering of thermal properties of silicon-based nanostructures. *Journal of Physics: Conference Series*, (IOP Publishing), 012086.
170. Yan, Z., G. Liu, J.M. Khan, and A.A. Balandin. 2012. Graphene quilts for thermal management of high-power GaN transistors. *Nature Communications* 3: 827.
171. Reina, A., et al. 2008. Large area, few-layer graphene films on arbitrary substrates by chemical vapor deposition. *Nano Letters* 9(1): 30–35.
172. Obraztsov, A.N. 2009. Chemical vapour deposition: making graphene on a large scale. *Nature Nanotechnology* 4(4): 212–213.
173. Subrina, S., Kotchetkov, D., and Balandin, A.A. 2009 Graphene heat spreaders for thermal management of nanoelectronic circuits. arXiv preprint arXiv:0910.1883.
174. Kim, H., A.A. Abdala, and C.W. Macosko. 2010. Graphene/polymer nanocomposites. *Macromolecules* 43(16): 6515–6530.
175. Li, X., et al. 2009. Large-area synthesis of high-quality and uniform graphene films on copper foils. *Science* 324(5932): 1312–1314.
176. Varshney, V., S.S. Patnaik, A.K. Roy, G. Froudakis, and B.L. Farmer. 2010. Modeling of thermal transport in pillared-graphene architectures. *ACS Nano* 4(2): 1153–1161.

ERRATUM TO

Chapter 7
Chip Size Packaging (CSP) for RF MEMS Devices

Li Xiao and Honglang Li

K. Kuang, R. Sturdivant (eds.), *RF and Microwave Microelectronics Packaging II*,
DOI 10.1007/978-3-319-51697-4_7

DOI 10.1007/978-3-319-51697-4_12

The original version of this book was inadvertently published with an incorrect author name in chapter 7. The correct author names are:

Li Xiao and Honglang Li

The updated original online version of this chapter can be found at
http://dx.doi.org/10.1007/978-3-319-51697-4_7

K. Kuang, R. Sturdivant (eds.), *RF and Microwave Microelectronics Packaging II*,
DOI 10.1007/978-3-319-51697-4_12

Index

K. Kuang, R. Sturdivant (eds.), *RF and Microwave Microelectronics Packaging II*,
DOI 10.1007/978-3-319-51697-4

Printed in the United States
By Bookmasters